THE NEW LANDSCAPE DECLARATION

THE NEW
LANDS

CAPE
DECLARATION

A CALL TO ACTION FOR THE TWENTY-FIRST CENTURY

RARE BIRD BOOKS   LANDSCAPE ARCHITECTURE FOUNDATION

THIS IS A GENUINE VIREO BOOK

A Vireo Book | Rare Bird Books
453 South Spring Street, Suite 302
Los Angeles, CA 90013
rarebirdbooks.com

A Vireo Book | Rare Bird Books Subsidiary Rights Department,
453 South Spring Street, Suite 302, Los Angeles, CA 90013.

Printed in the USA at GLS Companies

10 9 8 7 6 5 4 3 2 1

Publisher's Cataloging-in-Publication data:

Names: Landscape Architecture Foundation, editor.
Title: The New landscape declaration : a call to action for the twenty-first
century / edited by Landscape Architecture Foundation.
Description: Includes bibliographical references. | First Hardcover Edition |
A Genuine Vireo Book | New York, NY; Los Angeles, CA: Rare Bird Books, 2018.
Identifiers: ISBN 9781945572692
Subjects: LCSH Landscape architecture. | Landscape design. | Landscape
design—21st century. | BISAC ARCHITECTURE / Landscape
Classification: LCC SB472.45 N49 2018 | DDC 712—dc23

NEW LANDSCAPE DECLARATION SUMMIT AND PUBLICATION PROJECT TEAM

Landscape Architecture Foundation

Barbara Deutsch
Chief Executive Officer

Heather Whitlow
Director of Programs and Communications

Rachel Booher
Operations Manager

Dena Kennett
Summit Project Manager

Arianna Koudounas
Program Manager

Jennifer Low
Program Manager

Christina Sanders
Development Manager

Editor

Gayle Berens
Gayle Berens Consulting

Design Adviser

Norm Lee
Design Tower

Summit Graphic Identity

Andrew Schmidt
EDSA

Summit Event Planning Support

Krista Reimer

Summit Transcriptions

G. Adjoa Akofio-Sowah

The summit was also supported by over 20 volunteers.

ABOUT THE LANDSCAPE ARCHITECTURE FOUNDATION

Founded in 1966, the Landscape Architecture Foundation (LAF) is a 501(c)(3) nonprofit organization based in Washington, DC, with the mission to support the preservation, improvement, and enhancement of the environment. LAF advances the knowledge and influence of the landscape architecture profession through research, scholarships, and leadership initiatives.

ABOUT THE NEW LANDSCAPE DECLARATION SUMMIT

In June 2016, LAF held a summit featuring leading landscape architects from around the world. Building from LAF's *1966 Declaration of Concern*, which decried the burgeoning environmental crisis and heralded landscape architecture as a profession critical to help solve it, summit participants reflected on the last 50 years and shared aspirations and ideas for the future, which were used to craft a *New Landscape Declaration* for landscape architecture in the twenty-first century.

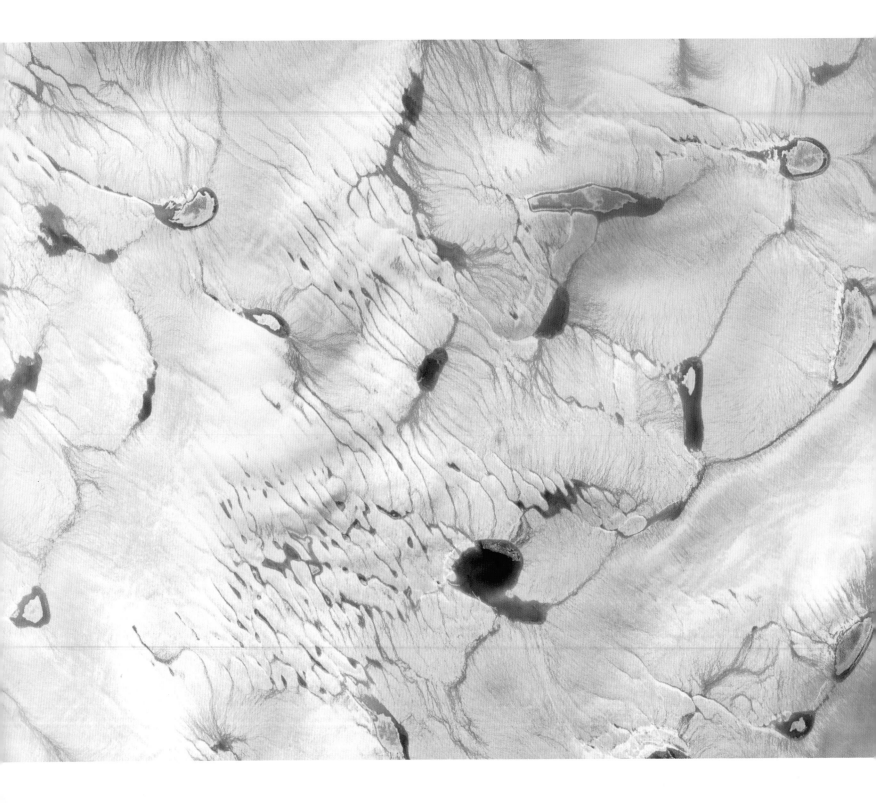

CONTRIBUTORS

Diane Jones Allen
DesignJones LLC

José Almiñana
Andropogon Associates

Gerdo P. Aquino
SWA
University of Southern
California

Thomas Balsley
Thomas Balsley Associates

Julie Bargmann
University of Virginia
D.I.R.T. studio

Henri Bava
Agence TER
Karlsruhe Institute of
Technology

Alan M. Berger
Massachusetts Institute of
Technology

Anita Berrizbeitia
Harvard Graduate School of
Design

Charles A. Birnbaum
The Cultural Landscape
Foundation

Keith Bowers
Biohabitats
Ecological Restoration and
Management, Inc.

Jacky Bowring
Lincoln University

Joe Brown
Planning and Design
Consulting Adviser to
AECOM

Nina Chase
Riverlife

Nette Compton
Trust for Public Land

Claude Cormier
Claude Cormier + Associés

James Corner
James Corner Field
Operations
University of Pennsylvania

Azzurra Cox
GGN

Julia Czerniak
Syracuse University

M. Elen Deming
University of Illinois at
Urbana-Champaign

Barbara Deutsch
Landscape Architecture
Foundation

Tim Duggan
Phronesis

Martha Fajardo
Grupo Verde Ltda

Mark A. Focht
New York City Parks and
Recreation

Gina Ford
Sasaki Associates

Christian Gabriel
US General Services
Administration

Ed Garza
Zane Garway

Christophe Girot
ETH Zurich

Maria Goula
Cornell University

David Gouverneur
University of Pennsylvania

Adam Greenspan
PWP Landscape
Architecture

Debra Guenther
Mithun

Kathryn Gustafson
GGN
Gustafson Porter

Feng Han
Tongji University

Andrea Hansen
Stae
Fluxscape

Susan Herrington
University of British Columbia

Randolph T. Hester
Center for Ecological Democracy
University of California, Berkeley

Kristina Hill
University of California, Berkeley

Alison B. Hirsch
University of Southern California
foreground design agency

Jeffrey Hou
University of Washington

Scott Irvine
University of Manitoba

Mark W. Johnson
Civitas

Joanna Karaman
OLIN

Mikyoung Kim
Mikyoung Kim Design
Rhode Island School of Design

Mia Lehrer
Mia Lehrer + Associates

Nina-Marie Lister
Ryerson University
PLANDFORM

Christopher Marcinkoski
University of Pennsylvania
PORT

Liat Margolis
University of Toronto

Deborah Marton
New York Restoration Project

Anuradha Mathur
University of Pennsylvania

Adrian McGregor
McGregor Coxall

Karen M'Closkey
University of Pennsylvania
PEG office of landscape + architecture

Blaine Merker
Gehl Studio

Elizabeth K. Meyer
University of Virginia

Brett Milligan
University of California, Davis
Metamorphic Landscapes

Tim Mollette-Parks
Mithun

Kathryn Moore
Birmingham City University

Alpa Nawre
Kansas State University
Alpa Nawre Design

Ellen Neises
RANGE
University of Pennsylvania

Cornelia Hahn Oberlander

Patricia M. O'Donnell
Heritage Landscapes LLC

Laurie D. Olin
OLIN
University of Pennsylvania

Kate Orff
SCAPE
Columbia University

Raquel Peñalosa
Raquel Peñalosa Architectes du Paysage

John Peterson
Harvard Graduate School of Design

Patrick L. Phillips
Urban Land Institute

Sarah Primeau
space2place design inc

Chris Reed
Stoss
Harvard Graduate School of Design

Stephanie Rolley
Kansas State University

Mario Schjetnan
Grupo de Diseño Urbano / GDU

Martha Schwartz
Martha Schwartz Partners
Harvard Graduate School of Design

Kelly Shannon
University of Southern
California

Dirk Sijmons
H+N+S Landscape
Architects

Ken Smith
Ken Smith Workshop

Laura Solano
Michael Van Valkenburgh
Associates

Nancy C. Somerville
American Society of
Landscape Architects

Anne Whiston Spirn
Massachusetts Institute of
Technology

Frederick R. Steiner
University of Texas at Austin

Carl Steinitz
Harvard Graduate School of
Design

Antje Stokman
University of Stuttgart

Marc Treib
University of California,
Berkeley

Charles Waldheim
Harvard Graduate School of
Design

Peter Walker
PWP Landscape
Architecture

Richard Weller
University of Pennsylvania

Marcel Wilson
Bionic

Kongjian Yu
Peking University
Turenscape

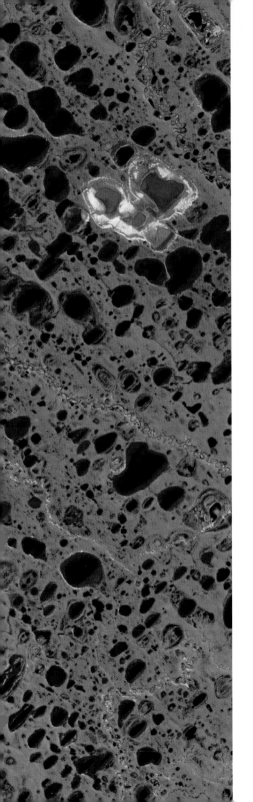

THE EARTH IS OUR CLIENT.

—GRANT JONES

CONTENTS

PREFACE

In June 2016, the Landscape Architecture Foundation (LAF) convened 75 preeminent thinkers and influencers, as well as over 600 attendees from around the world, to look at how landscape architecture can make its vital contribution to help solve the defining issues of our time. The New Landscape Declaration: A Summit on Landscape Architecture and the Future, held at the University of Pennsylvania, marked 50 years since Ian McHarg, Grady Clay, and other leading landscape architects composed LAF's seminal *1966 Declaration of Concern*, a call to action for landscape architecture to respond to what they called the "environmental crisis." Their declaration underscored the need for collaborative solutions and outlined a four-pronged strategy to multiply the effectiveness of the limited number of landscape architects because the need for better resource planning and design far outweighed the number of trained professionals.

In preparation for the summit, LAF asked a diverse group of leading minds in contemporary landscape architecture to write their own declarations, reflecting on the 1966 declaration and landscape architecture's achievements since then and offering bold ideas for what it should achieve in the future. These new declarations were presented at the summit on the first day. On the second day, nine panels, curated by theme, responded to the declarations and engaged in lively debate about how to realize landscape architecture's potential and effect real world change. This book presents those declarations in original essay form and showcases key ideas from the panel discussions.

LAF then synthesized the values, discussions, and ideas from the summit to create the *New Landscape Declaration*, this century's manifesto for the landscape architecture discipline. This poetically written, 400-word piece asserts the essential role of landscape architecture in solving the defining issues of our time: climate change, species extinction, rapid urbanization, and inequity. The recommendations are relevant to designers across the globe, underscoring the need to diversify, innovate, and create a bold culture of leadership, advocacy, and activism.

The summit and the *New Landscape Declaration* represented a unique opportunity for landscape architects to step back from their day-to-day work, come together under one tent, affirm values, and think big.

With this book, we invite you to engage in this critical, provocative, and inspirational examination of the power of landscape architecture, and to join the passionate community working to answer the call to action in the *New Landscape Declaration* at this critical time when the talents and services of landscape architects are so vitally needed.

This entire effort would not have been possible without the vision and tireless efforts of the Landscape Architecture Foundation Board of Directors and staff, in particular our 2016 Summit Task Force. From concept to execution, Richard Weller was a driving force. Dean Marilyn Jordan Taylor and PennDesign provided tremendous support in hosting the summit. We were blown away by the speakers and attendees who believed in the power of this endeavor and traveled from as far away as China, Argentina, and Australia to attend. LAF sincerely thanks all who contributed their ideas and passion to the process. We are proud, delighted, and humbled to have led this powerful and timely effort.

Barbara Deutsch   *July 2017*
Chief Executive Officer
Landscape Architecture Foundation

CHRONOLOGY

1828 *Landscape architecture* term first used

1863 *Landscape architect* first used as a professional title by Frederick Law Olmsted

1899 American Society of Landscape Architects founded

1900 Harvard University offers first degree in landscape architecture

1910 National examining board created for landscape architecture

1916 US National Park Service created

1927 World population reaches two billion

1949 *A Sand County Almanac*, Aldo Leopold

1960 World population reaches three billion

1961 *The Death and Life of Great American Cities*, Jane Jacobs

1962 *Silent Spring*, Rachel Carson

1964 Olmsted Centennial Exhibition

1964 US Civil Rights Act

1965 US Highway Beautification Act

1965 US White House Conference on the Environment

1966 Landscape Architecture Foundation incorporated (as the ASLA Foundation)

1966 LAF issues *Declaration of Concern*

1967 NASA releases whole Earth image

1969 *Design with Nature*, Ian McHarg

1970 First Earth Day

1971 *Life Between Buildings*, Jan Gehl

1970 US Clean Air Act

1972 US Clean Water Act

1972 United Nations Conference on the Human Environment

1973 US Endangered Species Act

1974 World population reaches four billion

1980 US Comprehensive Environmental Response, Compensation, and Liability Act (CERCLA, or Superfund)

1980 *The Social Life of Small Urban Spaces*, William "Holly" Whyte

1981 *Ecosystem services* term introduced

1984 E. O. Wilson introduces biophilia hypothesis

1985 *The Granite Garden: Urban Nature and Human Design*, Anne Whiston Spirn

1986 *Landscape Ecology*, Richard T. T. Forman and Michel Godron

1987 World population reaches five billion

1998 NASA scientist James Hansen delivers climate change warning to US Congress

1990 US National Environmental Education Act

1992 United Nations Earth Summit in Rio de Janeiro

1993 US Green Building Council established

1997 Kyoto Protocol limiting greenhouse gas emissions

1999 World population reaches six billion

2000 Paul J. Crutzen popularizes the term *Anthropocene* for the current geological epoch

2005 Millennium Ecosystem Assessment Reports released

2006 *The Landscape Urbanism Reader*, Charles Waldheim, ed.

2001 Al Gore begins giving his global warming slideshow

2011 World population reaches seven billion

2014 E. O. Wilson proposes designating half of the Earth's land as a human-free natural reserve

2015 Paris Agreement limits global warming to below two degrees Celsius

2016 LAF releases *New Landscape Declaration*

2016 United Nations Habitat III global summit on sustainable urbanization

2016 *Nature and Cities*, Frederick R. Steiner *et al.*

2016 Hottest year on record, third year in a row

A DECLARATION OF CONCERN

On June 1 and 2, 1966, at Independence Hall in Philadelphia, a small group of landscape architects who shared a concern for the quality of the American environment and its future were assembled by the Landscape Architecture Foundation (LAF). This was their declaration.

We urge a new collaborative effort to improve the American environment and to train a new generation of Americans equipped by education, inspiring example and improved organizations to help create that environment.

A sense of crisis has brought us together. What is merely offensive or disturbing today threatens life itself tomorrow. We are concerned over misuse of the environment and development, which has lost all contact with the basic processes of nature. Lake Erie is becoming septic, New York City is short of water, the Delaware River is infused with salt, the Potomac River with sewage and silt. Air is polluted in major cities and their citizens breathe and see with difficulty. Most urban Americans are being separated from visual and physical contact with nature in any form.

All too soon, life in such polluted environments will be the national human experience.

There is no "single solution," but groups of solutions carefully related one to another. There is no one-shot cure, nor single-purpose panacea, but the need for collaborative solutions. A key to solving the environmental crisis comes from the field of landscape architecture, a profession dealing with the interdependence of environmental processes.

Man is not free of nature's demands but becomes more dependent upon nature. Natural resources are where they are—not where we wish them to be. Those who plan for the future must understand natural resources and processes. These are the basis of life and the prerequisite for planning the good life. They must know geology, physiography, climatology, and ecology to know why the world's physical features are where they are, and why plants, animals, and man flourish in some places and not in others. Once they understand landscape capabilities—the "where" and "why" of environment, the determinants of change—they can then interpret the landscape correctly. Only then are they qualified to plan and design the environment.

Like the architect, the landscape architect practices a historic art. However, the landscape architect is uniquely rooted in the natural sciences. He is essential in maintaining the vital connection between man and nature.

The demand for better resource planning and design is expanding. Today's demands require far more landscape architects than are available. Schools are expanding, as are the ranks of practitioners, but they are stretched thin. The gap between demand and supply widens. The environment is being built hastily and too often without such professional advice or help. In the process, far too much is damaged beyond recall.

The solution of the environmental crisis demands the skills of many professions. So that the landscape architects may make their vital contribution, we propose a four-point program to bridge the gap between knowledge and practice: (1) recruitment, (2) education, (3) research, and (4) a nationwide system for communicating the results of research, example, and good practice. Its purpose is to multiply the effectiveness of the limited number of landscape architects while producing more trained people to cope with the future environment.

We pledge our services. We seek help from those who share our concern.

Campbell Miller

Grady Clay

Ian L. McHarg

Charles R. Hammond

George E. Patton

John O. Simonds

THE NEW LANDSCAPE DECLARATION

On June 10 and 11, 2016, over 700 landscape architects with a shared concern for the future were assembled by the Landscape Architecture Foundation (LAF) at the University of Pennsylvania in Philadelphia. Inspired by LAF's 1966 Declaration of Concern, we crafted a new vision for landscape architecture for the twenty-first century.

This is our call to action.

Across borders and beyond walls, from city centers to the last wilderness, humanity's common ground is the landscape itself. Food, water, oxygen—everything that sustains us comes from and returns to the landscape. What we do to our landscapes we ultimately do to ourselves. The profession charged with designing this common ground is landscape architecture.

After centuries of mistakenly believing we could exploit nature without consequence, we have now entered an age of extreme climate change marked by rising seas, resource depletion, desertification, and unprecedented rates of species extinction. Set against the global phenomena of accelerating consumption, urbanization, and inequity, these influences disproportionately affect the poor and will impact everyone, everywhere.

Simultaneously, there is profound hope for the future. As we begin to understand the true complexity and holistic nature of the earth system and as we begin to appreciate humanity's role as integral to its stability and productivity, we can build a new identity for society as a constructive part of nature.

The urgent challenge before us is to redesign our communities in the context of their bioregional landscapes, enabling them to adapt to climate change and mitigate its root causes. As designers versed in both environmental and cultural systems, landscape architects are uniquely positioned to bring related professions together into new alliances to address complex social and ecological problems. Landscape architects bring different and often competing interests together so as to give artistic physical form and integrated function to the ideals of equity, sustainability, resiliency, and democracy.

As landscape architects, we vow to create places that serve the higher purpose of social and ecological justice for all peoples and all species. We vow to create places that nourish our deepest needs for communion with the natural world and with one another. We vow to serve the health and well-being of all communities.

To fulfill these promises, we will work to strengthen and diversify our global capacity as a profession. We will work to cultivate a bold culture of inclusive leadership, advocacy, and activism in our ranks. We will work to raise awareness of landscape architecture's vital contribution. We will work to support research and champion new practices that result in design innovation and policy transformation.

We pledge our services. We seek commitment and action from those who share our concern.

PART I

INTRODUCTION

ACROSS BORDERS AND BEYOND WALLS, FROM CITY CENTERS TO THE LAST WILDERNESS, **HUMANITY'S COMMON GROUND IS THE LANDSCAPE ITSELF.** FOOD, WATER, OXYGEN—EVERYTHING THAT SUSTAINS US COMES FROM AND RETURNS TO THE LANDSCAPE. WHAT WE DO TO OUR LANDSCAPES WE ULTIMATELY DO TO OURSELVES. THE PROFESSION CHARGED WITH DESIGNING THIS COMMON GROUND IS LANDSCAPE ARCHITECTURE.

FROM THE NEW LANDSCAPE DECLARATION

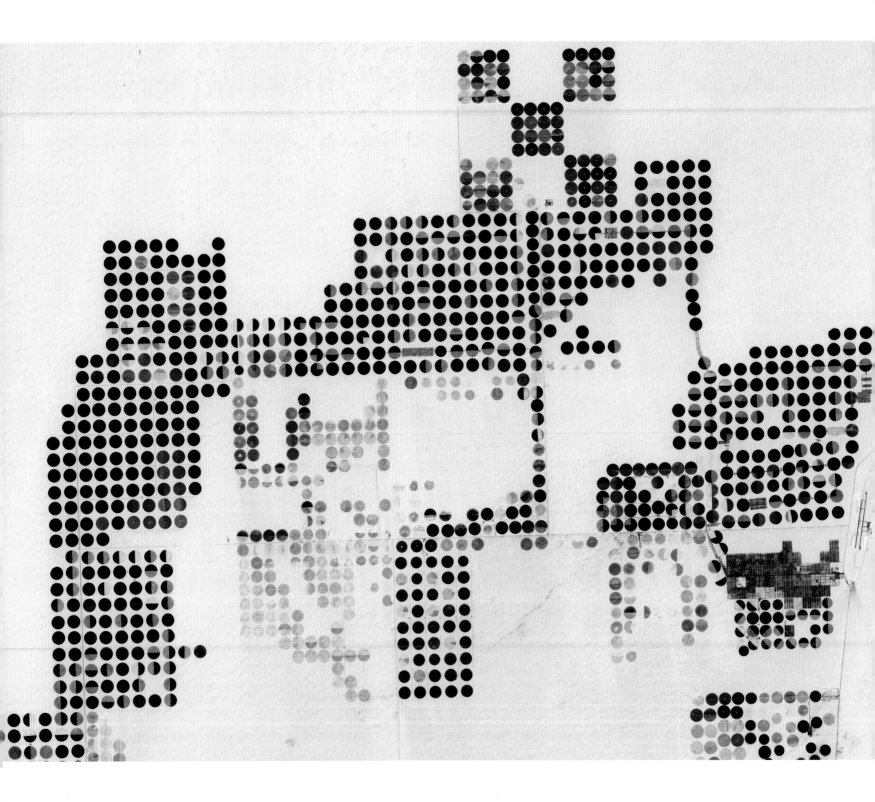

Chapter 1

OUR TIME?

RICHARD WELLER

In the Architectural Archives of the University of Pennsylvania, you can find the diaries of Ian McHarg. Penciled in for the day of June 2, 1966, McHarg had only one thing to do: meet with his contemporaries who had been assembled by the Landscape Architecture Foundation (LAF)—Grady Clay, Charles Hammond, Campbell Miller, George Patton, and John Simonds. Legend has it that the meeting took place at Independence Hall in Philadelphia where they drafted and read out a 490-word *Declaration of Concern*. They decried pollution, stressed that "man is not free of nature's demands," and hailed landscape architecture as "a key to solving the environmental crisis." They insisted that because of their basis in the natural sciences, landscape architects can interpret the landscape "correctly," and that therefore they are "qualified to plan and design the environment."

No one knows who really wrote the declaration or what the process was leading up to its publication, but the declaration's emphasis on understanding the landscape through its biophysical layers, not to mention its tone of bravado, would suggest that it was McHarg who held the pen.

The substance of the declaration was hardly earth-shattering, and there is no evidence that it attracted any media attention, but it did come at a significant moment in time. The year 1966 was bracketed by two hugely symbolic events. The first, in 1965, was the death of McHarg's alter ego—the charismatic champion of utopian modernism, Le Corbusier. The second, in 1967, was NASA's public release of the first whole Earth image. From here on, humanity would begin to comprehend its planetary ecological limits. Simultaneously, guided by Jane Jacobs, the design and planning professions began their paradigmatic shift to a concern for real people in real places. Then, three years after the seeds of the declaration had been sown, McHarg's magnum opus, *Design with Nature*, emerged fully formed, and it remains one of landscape architecture's most important books to this day.

The 32 declarations presented at the summit are now gathered in this commemorative volume and served as the raw material from which the final wording of the *New Landscape Declaration* was hewn.

In *Design with Nature*, with the entire planet as his stage and a dark city as his backdrop, McHarg repeatedly refers to landscape architects as stewards of the biosphere. It is here, with such grandiose pronouncements, that the global profession of landscape architecture would find both its (post) modern *raison d'être* and the impossibility of its realization. And it is this contradictory condition, in equal measure humbling and hubristic, that resonates through the profession and its academies to this day. It is what makes landscape architecture so compelling and, in a world so thoroughly changed by human hands, so pertinent.

The *1966 Declaration of* Concern had its limits. It was authored by five white men and focused on North America with no mention of equity, extinction, or climate change. As the fiftieth anniversary of the inception of the Landscape Architecture Foundation approached, it became obvious to the LAF leadership that the *1966 Declaration of Concern* required

renewal. This gave rise to a gathering, The New Landscape Declaration: A Summit on Landscape Architecture and the Future, held at the University of Pennsylvania on June 10 and 11, 2016. The summit provided the platform for a representative sample of selected landscape architecture academics and professionals to make new declarations and engage in debate with over 600 attendees. The 32 declarations presented at the summit are now gathered in this commemorative volume and served as the raw material from which the final wording of the *New Landscape Declaration* was hewn. These documents, alongside the new declaration, provide us with a historical opportunity to survey the profession's state of mind and speculate on its future. To that end, as I read them, three big topics emerged around which the 32 declarations orbit: climate change, urbanization, and the profession's identity in the twenty-first century.

Climate change

As a paleontological and contemporary phenomenon, climate change was known in the 1960s when the *1966 Declaration of Concern* was drafted but did not gain popular currency until two significant events: in 1979 when a National Academy of Sciences committee forecast temperature rise, and in 1988 when the United Nations Intergovernmental Panel on Climate

Change (IPCC) was formed. Rather than reading its absence from the 1966 declaration as shortsightedness, the original declaration's essential message—that human systems should be tuned to the earth's systems—is one that climate change makes more prescient, not less.

It is no mistake then that an overwhelming majority of the 32 new declarations refer to climate change and its symptoms as matters of priority. But apart from just using the expression to galvanize the profession, we need to ask what climate change really means for landscape architecture. Naomi Klein provides a big clue when she writes: "...climate change changes everything."

The climate has always been changing, so, technically speaking, climate change is nothing new, but anthropogenic climate change is different. We have now irrevocably altered natural history on a planetary scale, and while we have always made an impact, we have never before altered the fundamental workings of the earth system as a whole. Once you actually understand what this means, it is shocking. This is our Copernican revolution. What nature is or is not, and what it means to be human, are profoundly destabilized. In this sense, climate change is another name for, and evidence of, the arrival of the Anthropocene: an epoch in which cultural and natural histories have collapsed into one another and their fates rendered mutual. If anything beneficial is to come from our pumping billions of tons of carbon into the earth system, it is that its empirical existence ends the long history of mistaking nature as a mere resource we can exploit without consequence or venerating nature as something inviolable.

Nature in the era of climate change and the Anthropocene is then quite different from the *nature* invoked by the *1966 Declaration of Concern*. That 1960s nature was still a pure thing "out there," something being polluted, something to be saved, and for McHarg, a template we ought to study with scientific accuracy and then "correctly" follow. To design with the nature of the Anthropocene is, however, not so simple. Not only is nature now wildly unpredictable, it is also widely recognized as a cultural construct. This represents a shift away from the supposed harmony of sustainability that has dominated environmentalism for the latter half of the twentieth century, to the mutability of resilience.

Nature in the Anthropocene, the nature manifested by climate change, is not yet well known, but one thing is certain—it is now what we make it. And what we make it follows on directly from how we conceptualize it.

As was declared at the summit: engineers led the nineteenth century, architects the twentieth, and this is now our time.

And this is precisely what landscape architects do: whether we are aware of it or not, we give form to certain conceptualizations of nature. Our projects are, as it were, little worlds—experiments and case studies in synthesizing nature and culture in evermore ingenious and complex ways. Thus, we find ourselves in a historically and culturally significant position. Indeed, as was declared at the summit: engineers led the nineteenth century, architects the twentieth, and this is now our time.

But just saying it is our time does not make it so, and if there is one thing that almost all the new declarations glossed

> In the case of increasing density, designing high-performance public space—an art landscape architects have mastered over the past 50 years—will be critical to any city's future economic, social, and ecological quality of life.

over, it is that landscape architecture still lacks the self-critical philosophical underpinnings that are needed to restrain its messianic tendencies and make more credible its claims to large-scale land use planning and urban design, let alone planetary stewardship. Maybe landscape architecture is not yet big enough for criticism and we must band together to build the profession, but as other professions and disciplines have demonstrated, it is ultimately criticality, not backslapping, that forges a profession that the public looks up to.

Urbanization

In tandem with changing the chemical composition of the atmosphere, over the last 50 years humanity has also altered the surface of the planet with urbanization and its related infrastructure as never before. This historical phenomenon seems likely to continue for much, if not all, of the twenty-first century as world population moves into double-digit billions. If birthrates and migration from rural to urban areas continue to increase as demographers expect them to, then we can reasonably assume that an additional three billion people will become urbanized between now and 2100. The equivalent of over 350 New York Cities will be needed to accommodate them—a little over 4.2 constructed each year.

Around the world, urban growth is occurring as both informal and planned development and pushing in two directions—centrifugal sprawl on the one hand and centripetal densification on the other. As many of the declarations attest, landscape architects have leading roles to play in these processes. In the case of peri-urban growth, getting involved upstream of the development process would give landscape architects the opportunity to direct the growth, interweaving it with agricultural lands and remnant habitat. In the case of increasing density, designing high-performance public space—an art landscape architects have mastered over the past 50 years—will be critical to any city's future economic, social, and ecological quality of life.

Whereas McHarg, quoting architect Peter Blake, referred to the modern city as "God's own junkyard," over the course of the past 50 years—as the tendrils of urbanization have become ubiquitous—design and planning professions have come to see the city not as the problem but as a crucial part of the solution. While it may sound like spin, it is more than that. The city is a clustering of commerce, culture, architecture, and technology that has not changed much in 10,000 years but is now being fundamentally reconceptualized as something that could be continuous with, instead of just blithely resistant to, ecological flows. This reconceptualization opens the way for the practical and innovative redesign of urban systems *and* the global supply chains they depend upon.

Converting cities from industrial machines to ecological systems is no simple thing, but as ecologists, engineers, architects, planners, and developers—along with the general public—begin to think of cities as a new kind of nature rather than something opposed to nature, landscape architects find themselves at a propitious moment in time. Our propensity for holistic thinking and interdisciplinary collaboration, and our grasp of the systemic, relational, and temporal nature of things, along with the increasing sophistication of available data, mean landscape architects are well situated to participate in, if not lead, this urban transformation. Not only will cities become increasingly sophisticated ecological systems, but so too can we speculate that sometime in the twenty-second century, when world population numbers decline, agricultural production and food distribution will be more efficient and large-scale landscape restoration will be undertaken.

At least that is the theory, and the hope. We must first get through this century's bottlenecks with some semblance of the ecosystem intact. This challenge is mired in spiritual, political, and economic orthodoxies that resist change, but it is also a design challenge. And, as many of the declarations stress, the profession of landscape architecture not only

> Cities are now being fundamentally reconceptualized as something that could be continuous with, instead of just blithely resistant to, ecological flows.

feels a sense of responsibility, it also has the ability to make a constructive difference in parts of the world where the pressures of climate change and resource depletion are being felt with particular acuity.

Professional identity

The problem for the profession, however, is that these pressures are shaping territory where landscape architecture has very little capacity. As we move out from the wealthy enclaves of the developed world and follow their supply chains to the frontiers of extinction, extraction, and waste, landscape architecture's influence diminishes every step of the way. Taking this opportunity to reflect on the profession's identity and scope, we must stare into the chasm between the many things landscape architects *say* they could do and what they *actually* do.

To take the large-scale landscape issues of the Anthropocene seriously suggests a need for the significant expansion of landscape architecture's professional and educational capacity, something the *1966 Declaration of Concern* called for 50 years ago. This suggests the need for the profession's representative bodies, along with its educational programs and its practices, to ask themselves some questions: How can landscape architecture build capacity around the world rather than just export commercial services? How can work be created rather than just received? What knowledge is needed? What methods are most suitable? Whose interests will the results really serve?

How can landscape architecture build capacity around the world rather than just export commercial services? How can work be created rather than just received? What knowledge is needed? What methods are most suitable? Whose interests will the results really serve?

As I write this, I am keenly aware that institutions, schools, and practices can hardly countenance expansion when they are struggling just to hold their ground, but if the profession is to close the gap between what it says and what it does, then individuals and organizations need to be more ambitious and more adventurous. It also means that landscape architecture has to be better at communicating its global potential. But this cannot just be grandiloquence: it must be built on research and design projects of substance.

Many of the declarations champion the potential of new modes of practice and new constituencies, right here in the heart of the developed world. Our cities are riven by issues of social and ecological justice, which landscape architects can either disguise or confront in their work. For example, the agency of public space in retrofitting shrinking cities and the emergence of so-called *nature-based* strategies for coastal resilience have recently opened up challenging and rewarding roles for the landscape architect as the curator of sociopolitical and ecological processes as much as the provider of amenities.

In reading the declarations, there can be no mistake that the twenty-first-century landscape architectural project is one of social justice, ecological synthesis, and territorial reach, but some declarations remind us that good design manifests civility and that, at its best, the language of landscape is poetic. One of the panel discussions at the summit was devoted to questions of aesthetics, and in all the talk of problems and solutions, its message that design is akin to art rang clear. Indeed, for a field that claims the extraordinary history of gardens as its artistic legacy—not to mention aspects of modernism, earth art, and contemporary digital imaging—there is surprisingly little discussion or criticism of what contemporary landscape aesthetics are and what they might yet become.

The relative absence of discussions of aesthetics is in part landscape's great conceit: because it often looks benign, familiar, or 'natural,' no one thinks it is saying anything. The authors of contemporary landscapes need to worry about this issue. Irrespective of whether the images are true or false, or somewhere in-between, imagining its landscape is one of the primary ways that culture makes sense of its time and place in history and by which it creates and contests hegemonic meaning. As elemental as it is, the ecological crisis is also a crisis of meaning. Simply put, nature in the Anthropocene cannot look like the nature of the Holocene.

We can now ask how this new landscape declaration can help our small, relatively powerless, and yet critically important, profession make its "vital contribution." In essence, the *New Landscape Declaration* calls upon students, practitioners, and academics to work to diversify and expand the profession through "inclusive leadership, advocacy, and activism." It asks us to reflect on how we can reach toward the ideals of "equity, sustainability, resiliency, and democracy." It asks how we can help "create places that serve the higher purpose of social and ecological justice for all peoples and all species." It asks how our designs "nourish our deepest needs for communion with the natural world and with one another." It asks how they "serve the health and well-being of all communities."

We are asking a lot of our profession. Indeed, most of us fall well short of meeting those ideals. But that is surely the point: ideals are beacons, not ends. In this way, the *New Landscape Declaration* is not only about the profession as an anonymous whole, it is also a call for personal reflection on what it means to be a landscape architect at this moment in history and that might be where its words take real effect. It is from there that inspired and authentic action can emerge and gather momentum.

Ideals are beacons, not ends. In this way, the *New Landscape Declaration* is not only about the profession as an anonymous whole, it is also a call for personal reflection.

By what they say, and in part by what they do not say, the 32 declarations collected here are good to think with. They speak to an uncertain new world of climate change and global urbanization but do so with clear-eyed confidence in the profession's values, abilities, and potential. Each declaration has some wisdom that will help you form your own answers to the challenges the new declaration presents. This is not a time to be cynical: let the *New Landscape Declaration* be our Hippocratic oath.

Richard Weller is Martin and Margy Meyerson Chair of Urbanism and professor and chair of landscape architecture at the University of Pennsylvania School of Design where he teaches history, theory, and advanced design studios. He received his BLA from the University of New South Wales in Sydney and his MLA from RMIT in Melbourne.

Chapter 2

TEXT AND CONTEXT: THE *1966 DECLARATION OF CONCERN*

ELIZABETH K. MEYER

The following is an edited version of remarks delivered at the summit and is not intended to be an essay or verbatim representation.

The *1966 Declaration of Concern* was written and signed by six men on behalf of our professional organization and foundation—Grady Clay, Charles R. Hammond, Ian McHarg, Campbell Miller, George Patton, and John Simonds. There is still much that is unknown about the event itself—who was involved, what their motivations were, who their audience was, and what the social and political context was within which they operated. But based on my preliminary research, I will assert that a sociopolitical conception of the "good life" and the role that urban nature and the constructed landscape play in the "good life" was at the heart of this early Landscape Architecture Foundation enterprise.

Professional context

In the 1960s, the landscape architecture profession in the United States was a very small group. The American Society of Landscape Architects (ASLA) had a few thousand members. There were only 15 to 20 university programs in landscape architecture. Nonetheless, this small profession was creating notable designed landscapes of artistic merit, including modernist work that is now canonical in our field. These projects responded to the challenges of urban living with new spatial and functional typologies in the public and private realms. The commissions crossed a wide array of scales

> Through these exhibitions and publications, landscape architects were...staking out the value of a landscape architecture approach in formulating a new urban ecological imaginary.

from small projects to those that took on the metropolis and included innovative regional watershed planning projects.

Looking only at the remarkable projects of the era might lead us to wonder what the crisis was that precipitated the *1966 Declaration of Concern*. But if we look at the influential books that were written in this period, such as Rachel Carson's *Silent Spring*, Lawrence Halprin's *Cities*, and Jane Jacobs' *The Death and Life of Great American Cities*, we understand that there was indeed a crisis—a concern about the rapidly changing urban landscape of the United States and the capacity of cities to absorb growth without compromising environmental health and human flourishing.

Concurrent with broader, if nascent, cultural awareness of the complex interconnections between urban patterns, environmental change, and human well-being, ASLA sponsored a major traveling retrospective on nineteenth-century landscape architect Frederick Law Olmsted. Curated by Professor Norman Newton of Harvard Graduate School

of Design and his students, this 1964 exhibition introduced Olmsted and, by extension, the landscape architecture profession, as an agent of urban social and environmental reform. Two years later, Julius Gy Fabos published *Frederick Law Olmsted, Sr.: Founder of Landscape Architecture in America*. Through these exhibitions and publications, landscape architects were connecting historical innovations in urban landscape design to contemporary urban environmental concerns. They were staking out the value of a landscape architecture approach in formulating a new urban ecological imaginary.

National urban policies and political context

In the mid-1960s, in many cities, the air and water pollution caused by the unregulated release of industrial waste was so intolerable that the federal government acted. The first Clean Water and Clear Air Acts were passed in the early 1960s, and by the decade's end, the National Environmental Protection Act (NEPA) was passed and the Environmental Protection Agency (EPA) was created.

At the same time, federal funding facilitated massive destruction of neighborhoods and dislocation of minority communities in the name of improvement, urban renewal, and highway construction. Urban rioters, peaceful and destructive, demanded social and political change in parallel with the environmental movement to improve air and water quality. Seminal civil rights legislation, such as the Voting Rights Act, followed the assassination of President John F. Kennedy as the public and legislators advocated for change. Vice President Lyndon Baines Johnson assumed the presidency during this unsettled time.

President Johnson and his wife, Lady Bird, were key actors within the political generation that wrote and passed groundbreaking civil rights and environmental legislation. Lady Bird Johnson, a committed environmentalist, had a large influence on the president. She was not just a beautifier who liked flowers, as she was often dismissively described; she was a tireless and effective advocate for urban parks, highway redesign, and the conservation of native and local plant communities.

To fully appreciate the import of the *1966 Declaration of Concern*, we need to look beyond what landscape architects were saying to each other and what they were building. We need to understand landscape architects' relationships with people outside the profession—their clients, public figures, and politicians. For instance, a year before the declaration was written, President Johnson was directly involved through his writings and programs in debates about the future of our cities and countryside. In his 1965 State of the Union and his 1964 Great Society speech at the University of Michigan, President Johnson identified natural beauty as a factor that sustained the American spirit and enlarged its vision. To save the countryside and enrich our cities, he called for the creation of more imaginative urban programs, better urban design, and natural resource conservation.

The president followed up these addresses with a special meeting, the White House Conference on Natural Beauty, in spring 1965. It was probably not a coincidence that several of the declaration authors—Grady Clay, Ian McHarg, and John Simonds—participated in that conference. Grady Clay was the chair of the Waterfronts Committee that included

> President Johnson...called for creating more imaginative urban programs to save the countryside and enrich our cities, for better urban design, as well as natural resource conservation.

National Park Service Director Conrad "Connie" Wirth and Christopher Tunnard. Ian McHarg was on the Landscape Action Committee with William "Holly" Whyte and Phil Lewis. John Simonds chaired Parks and Open Space with Jane Jacobs and Charles Eliot II. This conference was attended by 115 people, including senators, politicians, designers, and planners. The declaration signers were part of a larger interconnected network of very influential and caring people and the declaration evolved in response to this professional and political opportunity as well as to the decade's urban and environmental crises.

Lady Bird Johnson also believed that the experience of urban landscape, of natural beauty as she called it, was vital to vibrant, healthy cities. In January of 1965, she wrote in her diary: "Getting on the subject of beautification is like picking up a tangled skein of wool…all the threads are interwoven—recreation and pollution and mental health, and the crime

> Lady Bird Johnson was a significant actor, facilitator, and intermediator, connecting federal legislators and politicians—and their urban policies and environmental priorities—to the small but vital profession of landscape architecture in the 1960s.

rate, and rapid transit, and highway beautification, and the war on poverty, and parks—national, state, and local. It is hard to hitch the conversation to one straight line because everything leads to something else."

Lady Bird's conception of the landscape was much more complex than that of many landscape architects in the 1960s. It resonates with our early twenty-first-century conceptions of the landscape as a mesh of human and nonhuman life and matter. She was more than a beautification advocate; she was a patron of McHarg (Anacostia River master plan), Halprin (Potomac River watershed plan), and many other landscape architects. She knew political figures like Connie Wirth, Secretary of the Interior Stewart Udall, and Laurance Rockefeller, a generous environmental philanthropist who funded the White House Conference on Natural Beauty. And

she was described by McHarg in his autobiography, *A Quest for Life*, as a fan of his "who takes notes during my speeches, assuring me she would persuade the president to a more ecological viewpoint."

Lady Bird Johnson was a significant actor, facilitator, and intermediator, connecting federal legislators and politicians—and their urban policies and environmental priorities—to the small but vital profession of landscape architecture in the 1960s. Her voice and lifelong convictions are behind the correlations that President Johnson made between human flourishing and the experience of constructed and natural beauty.

While Lady Bird was not one of the signers of the *1966 Declaration of Concern*, all of them interacted with her in some way.

A year after the White House conference, the *1966 Declaration of Concern* was written in response to challenges as well as to an opportunity—a possibility of greater professional visibility. But they were also responding to a perceived threat, in that there were several references in *Landscape Architecture Magazine* in October 1964 to a 1963 American Institute of Architect's declaration that asserted that the architects' responsibility was "for nothing less than the nation's man-made environment, including the use of land, water, and air." In short order, ASLA formally responded in writing by rejecting AIA's action proposals, created their own foundation, had members participate in the White House conference, and through its foundation issued the *1966 Declaration of Concern* on the environment. All this was happening concurrent with the 1965-67 Potomac River watershed planning study—one of the first in the United States. That Secretary of the Interior Stewart Udall invited AIA and not ASLA to appoint an interdisciplinary team to undertake "a model study" of the Potomac River could not have been lost on task force members Grady Clay and Ian McHarg or the University of Pennsylvania landscape architecture graduate students who contributed most of the analytical maps, diagrams, and drawings to this remarkable report under the supervision of David Wallace and Narendra Juneja.

The *1966 Declaration of Concern* and its significance for the profession

The creation of LAF and the writing of the declaration may have been a response to what cultural geographer Denis Cosgrove would describe as "the tensions of the time" that influence how landscape architects conceive and create. The first tension was the everyday experience of cities—pollution, urban renewal, and sprawl—that threatened the immediate and long-term health of individual human beings within particular urban ecosystems. The second tension of the time involved the actions of both politicians and citizens as they solicited multidisciplinary professional expertise to propose new design and planning paradigms capable of analyzing these complex urban environmental problems and proposing new urban ecological imaginaries. And the third tension of the time was the perceived professional threat from the profession of architecture that was eager to expand its scope beyond the building envelop into an emerging field, urban environmental design, which had too few practitioners to do the immense work that was necessary to preserve and revitalize the American environment.

This assertion of the landscape's role in the American city connects the May 1965 White House conference to the *1966 Declaration of Concern*. The White House event was rife with statements such as those by Frederick Gutheim: "The American city has a spacious quality and at best incorporates a natural framework and landscape pattern that runs into almost every block." And John Simonds: "...it is not enough to build open spaces into our cities; we must rather conceive of each city as an interrelated park...." The declaration signers extended this to the need for analyzing a region's natural resources and processes as the basis for planning and design.

The *1966 Declaration of Concern* is a call to action for the socio-ecological city, the just city, and the good life. It is

> The *1966 Declaration of Concern* is a call to action for the socio-ecological city, the just city, and the good life.

not simply about solving environmental problems—it makes connections between environmental quality and the quality of human life. The "good life" does not just refer to Aristotle. Students of American history will know that LBJ's Great Society envisioned a society that fulfilled more than the needs of the body and the demands of commerce, but the desire for beauty and the hunger for community, to create a flourishing community where our people can come to live the "good life." The "good life" in the declaration requires beauty as well as function, aesthetics and landscape experience as well as

> Landscape architecture is based on a professional ethos that recognizes that designers construct human experiences, and that those experiences alter perceived and actual relationships between humans and the planet.

economics, within the urban environment. The "good life" reference in the declaration and the role of aesthetics, as well as ecological analysis in landscape architecture, are veiled to us if we do not understand LBJ's Great Society rhetoric.

The last part of the declaration is also important as it relates to the issue of a professional ethos and the scope of our work: "Like the architect, the landscape architect practices a historic art. However, the landscape architect is uniquely rooted in the natural sciences. He is essential in maintaining the vital connection between man and nature." This passage is all the more meaningful when we know about AIA's declaration that it is the architect who is responsible for the use of land, water, and air in the man-made environment. I would suggest that the declaration implies that landscape architects are responsible for something different and more socially profound—maintaining the vital connection between man and nature.

Landscape architecture is based on a professional ethos that recognizes that designers construct human experiences and that those experiences alter perceived and actual relationships between humans and the planet. It is the landscape architect that extends the rhetoric of the Great Society into a spatial practice. That practice does more than use natural resources wisely. It is predicated on shaping spaces for both human and nonhuman life and processes to flourish. Landscape architecture foregrounds the entanglement of the human and nonhuman beings and processes.

We could spend a lot of time celebrating the things that have happened because of, and after, the *1966 Declaration of Concern*. For example, the number of graduate and undergraduate landscape architecture programs in the United Stated doubled within six years.

I would suggest that this legacy involves a quest for the "good life," a life that requires more than bodily needs and the demands for commerce. The signers of the declaration could not have imagined the types and quality of practices that exist today, but we share many of their same concerns. Those of us who studied landscape architecture in the 1970s owe a debt to those who practiced in the 1960s, who participated in the White House Conference on Natural Beauty, who interacted with political leaders at the local and national levels, and who wrote the *1966 Declaration of Concern*. Today, there are 20 times the number of landscape architects in the United States as there were in the 1960s. We are the beneficiaries of their vision, audaciousness, and tenacity.

But their socio-ecological vision has yet to be realized. Today, visionaries of the "good life" are working at the local scale, in cities and communities. In our practices, we need to remember that 50 years ago, a small group of landscape architects declared their concerns not just for the environment; they staked out a vision of who had the right to the city, to use a phrase of Henri Lefebvre, and who had the right to flourish through their everyday experience of man and nature or the human and nonhuman, made palpable through design and planning of cities and settlements.

Elizabeth K. Meyer is Merrill D. Peterson Professor of Landscape Architecture and Dean, School of Architecture, University of Virginia. She holds a BS in landscape architecture and an MLA from the University of Virginia and a master of art in the history of architecture and urban development from Cornell University.

PART II

THE NEED FOR ACTION

AFTER CENTURIES OF MISTAKENLY BELIEVING WE COULD EXPLOIT NATURE WITHOUT CONSEQUENCE, **WE HAVE NOW ENTERED AN AGE OF EXTREME CLIMATE CHANGE** MARKED BY RISING SEAS, RESOURCE DEPLETION, DESERTIFICATION, AND UNPRECEDENTED RATES OF SPECIES EXTINCTION. SET AGAINST THE GLOBAL PHENOMENA OF ACCELERATING CONSUMPTION, URBANIZATION, AND INEQUITY, THESE INFLUENCES DISPROPORTIONATELY AFFECT THE POOR AND **WILL IMPACT EVERYONE, EVERYWHERE**.

FROM THE NEW LANDSCAPE DECLARATION

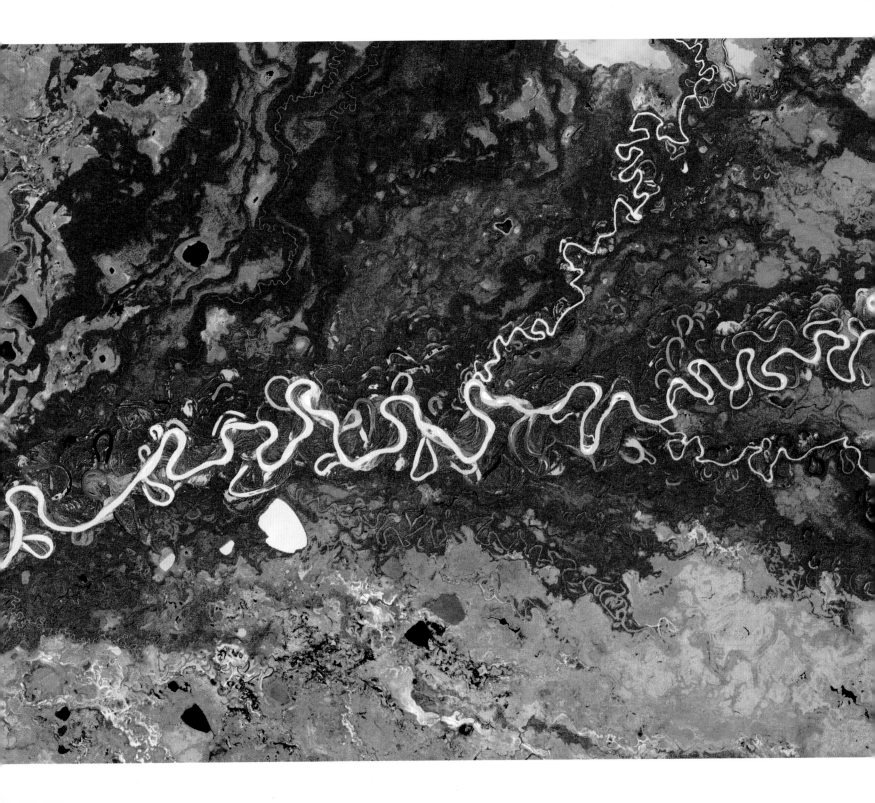

Chapter 3

BEYOND PRACTICE: LANDSCAPE ARCHITECTS AND THE GLOBAL ECO-CRISIS

MARTHA SCHWARTZ

The scientific consensus is that anthropogenic climate change is now indisputable. James Hansen, the leading climate scientist previously from the National Aeronautics and Space Administration (NASA) and respected for many predictions that have proven to be true, recently issued a grim warning that we are nearing the point of no return when it comes to reversing or even mitigating the adverse effects of climate change.

Several alarming facts provide evidence that global warming is upon us and happening faster than predicted: (1) 2015 was the warmest year on record, and this past decade is the warmest since 1880; (2) in November of 2015, a one-degree-Celsius planetary rise in temperature was officially acknowledged (but widely believed to be conservative)—the halfway mark to the two-degrees-Celsius target or safe limit to avoid catastrophic global warming; and (3) East Siberian Arctic Shelf (ESAS) methane is being released—the most threatening fact of all.

We have exceeded the projected tipping point of 350 parts per million of carbon dioxide (CO_2) and are now at 400 parts per million, heating up our land, air, ice, and oceans with the equivalent of 400,000 Hiroshima atomic bombs per day. Global ocean temperatures are now one degree Celsius higher than they were 140 years ago. The heated Arctic Ocean is causing the permafrost of the ESAS to melt, releasing methane—a gas 20 to 30 times more potent than CO_2 as a heat-trapping gas—into the atmosphere. There are such massive reserves of methane in the subsea Arctic that if only a small percentage is released, it can lead to a jump in the average temperature of the Earth's atmosphere by ten degrees Celsius.

Recent observations in the Siberian Arctic show that increased rates of methane are escaping from the seabed *now*. These facts have produced a plausible scientific prediction of a catastrophic release, or bubble, of methane occurring abruptly or in coming decades. Such a release could have an exponentially amplifying effect on global warming, launching

We must advocate for funding the development and testing of a portfolio of geotechnologies to counteract man-made global warming until the required reductions in CO_2 emissions are reached and we have transitioned to sustainable energy economies.

catastrophic scenarios more rapidly than we had anticipated.

The *1966 Declaration of Concern* correctly predicted and responded to the environmental crisis with a vision 50 years ago. The ecological planning initiatives and educational goals that were projected have been accomplished. At this point, however, we confront a drastic new challenge. In 1966, those visionaries could not foresee globalization or the population explosion with corresponding fossil fuel use and consumption that have outstripped all our profession's good intentions and achievements regarding sustainable development through responsible design.

I no longer believe that the work we do as individual, responsible practitioners will be able to contribute effectively to averting this long-predicted crisis because we are entering a state of emergency. We do not have another 50 years or perhaps even 15. Sadly, our excellent professional practices will become irrelevant in the face of global warming, a problem whose magnitude we are now confronting. I do not advocate putting an end to our professional excellence or our individual duties as responsible practitioners. But my message is: we must go beyond landscape architecture practice in order to broach this critical environmental crossroads.

The questions before us are: What can be done to keep this scenario from playing out? What can we do as a group of people whose ethos is to steward our natural environment, since the impacts are coming much sooner than expected?

My declaration is for a collective call to action. We must advocate for funding the development and testing of a portfolio of geotechnologies to counteract man-made global warming until the required reductions in CO_2 emissions are reached and we have transitioned to sustainable energy economies.

As a first priority, we should develop techniques to cool the Arctic because the possibility of a major methane release triggered by melting Arctic ice constitutes a planetary emergency. Technologies have been proposed for rapidly cooling the Arctic on the necessary scale in the form of solar radiation management (SRM). We should be investing in research and development toward this goal immediately.

At the same time, we must focus on measures that can reduce existing quantities of atmospheric CO_2 by carbon dioxide removal (CDR) processes to lower the pollutant level and warming effects.

Finally, cutting global greenhouse gas emissions must remain an urgent priority; thus, reducing emissions from existing, new, and proposed carbon power stations, particularly coal plants, with carbon capture and storage (CCS) techniques, must be rigorously pursued.

Scientists have conceived various methods—and new ones may be discovered—to achieve these goals, so it is very

As a first priority, we should develop techniques to cool the Arctic because the possibility of a major methane release triggered by melting Arctic ice constitutes a planetary emergency.

likely to be technically feasible. But much more research and testing is needed before deployment. I believe that science can help us out of this imminent and dire situation in order to buy time so that the longer-term goal of zero carbon emissions can eventually be achieved.

Therefore, I urge our professional organizations to create a political wing with a forceful agenda to persuade our decision makers and politicians to support bold research to save our planet's atmosphere through technologies that can prevent Arctic methane release, and sequester and capture carbon dioxide. We must exert pressure on our government to fund the equivalent of a "Manhattan Project for Climate Change Mitigation" in response to the clear and present environmental danger we are now facing, especially with regard to Arctic methane release. Such a political agenda should also have a focused social media voice. It is through these modalities that change can occur. Petitions and signatures impel those in power to exert the voice of the people. This is today's version of taking to the streets. We must become online warriors.

We are a well-educated, knowledgeable group that has the status to influence people. Together, we have the power to create awareness about this environmental emergency and to make change. I further propose that we, as a group of dedicated landscape architects, immediately embark upon a hyperaggressive climate campaign that should be owned by the two most important professional organizations that represent us: the American Society of Landscape Architects (ASLA) and the Landscape Architecture Foundation (LAF). We must urge these two organizations to officially recognize the gravity of the situation and align with and support other actionable networks, such as 350.org, Friends of the Earth, Greenpeace, the Arctic News, and the Arctic Methane Emergency Group, among others, that are working heroically and aggressively on climate change policy and action.

Finally, ASLA has a lobbying arm in Washington, DC. We must exercise our full intellectual authority and political influence to strategically advance climate rescue.

In summary, I am suggesting that we shift our priorities from individual practice to group political action so as to impel our government to forge an international effort to cool the Arctic, suppress methane, and aggressively remove CO_2 from the atmosphere; take immediate and extremely drastic action to entirely curb global carbon dioxide emissions; and evolve rapidly toward completely renewable energy resources.

I have hope that the world's best scientists will find ways to buy us the gift of time so that we can avert the swift intensification of climate change. Then we will be given a second chance to learn how to live in balance with the earth. But we must act together now.

The author would like to thank Edith Katz, formerly of Martha Schwartz Partners, for her invaluable assistance.

Martha Schwartz is principal of Martha Schwartz Partners and has over 35 years of experience as a landscape architect, urbanist, and artist. Schwartz is professor in practice of landscape architecture at Harvard Graduate School of Design, a founding member of the Working Group of Sustainable Cities at Harvard University, and was appointed in 2015 as a Built Environment Expert (BEE) of the Design Council Cabe.

Chapter 4

OF WILDERNESS, *WILD-NESS*, AND WILD THINGS

NINA-MARIE LISTER

And I think in this empty world there was room for me and a mountain lion.
And I think in the world beyond, how easily we might spare a million or two humans
And never miss them.
Yet what a gap in the world, the missing white-frost face of that slim yellow mountain lion!
—D. H. Lawrence, *Mountain Lion*

Humans are an urban species. For the first time in our history, more than half the world's 7.4 billion humans now live in urban settlements. We have become the single dominant species shaping the planet, from its surface lands and waters to its climate and, by extension, to the future of all other species on earth. The age of Anthropocene is upon us and we are its defining creature.

But what of the other 2.5 million species we know of so far (by the most conservative estimate)? Who in the Anthropocene will speak for these creatures and their wild places? Where will these wild things be and, through their fading reflection, what will become of the wild within the human?

In the last 20 years, landscape architecture has risen to prominence—and in some cases to dominance—within the applied professions of city building and urban placemaking. In North America, the most urgent challenges posed by the environmental crises of the mid-twentieth century (some of which are referred to in the landmark *1966 Declaration of Concern*) have been, to a large extent, recognized, managed, remediated, and, in a few rare cases, solved. Indeed, the rise of the 1970s and 1980s third-wave environmentalism was activated in large part through landscape architecture and supported by allied disciplines of ecology, environmental planning, environmental studies, and associated sciences. Together with landscape architects, these allies advocated, planned, and designed for finding environmentally responsible solutions, reducing and cleaning up toxic waste, controlling pollution, improving waste management, and initiating environmental conservation. These and other strategies were effective reactions as crisis management but have now given way to more proactive strategies for longer-term, larger-scale, complex challenges related to climate change and sustainability.

Landscape architecture has been at the center of this shift—from new urbanism to landscape urbanism to ecological urbanism—landing squarely in the rhetoric of resilience and the practice of green infrastructure. Some might conclude that the landscape architect has arrived, center stage, in the Anthropocene as urban savior. But on this urbanizing planet, what remains of the wild? More urgently, what will become of the wild things and their places and of the *quality of being* that defines them and, by contrast, us?

On the relentless trajectory of global urbanization, we continue to lose millions of acres each year of Earth's natural and agricultural cover through land conversion. The loss of natural habitats, whether by swift condemnation and conversion or by the cumulative paper cuts of habitat fragmentation and degradation, ultimately leads to irretrievable loss of biodiversity. The Anthropocene is the planet's sixth great extinction epoch: from almost daily extirpation to mass extinction, the wealth of the world's biodiversity is bleeding away. While we may lament the loss of the wild, we also exacerbate it by failing to validate and value what it is to *be* wild. Honoring the condition of *wild-ness* is fundamental to valuing the wild things and caring for their places—central tenets in activating their protection.

The wild and its essence will not persist if we retreat passively. We cannot simply do nothing, for neglect is *not* benign. A different wild will inevitably emerge from the void left behind: from invasive species to barren fields and hostile environments, an evolving new nature—an unintended consequence of our own design—will simply select humans out, replacing us with plague and pest alike. Our role must be as

> Honoring the condition of *wild-ness* is fundamental to valuing the wild things and caring for their places—central tenets in activating their protection.

active agents in reaffirming, reestablishing, and revaluing the place and role of the wild. Policies and targets for wilderness protection vary widely, from the United Nations Convention on Biological Diversity's goal of 17 percent by 2020 to ecologist E. O. Wilson's ambitious "Half-Earth" movement to protect 50 percent of the world's natural landscapes from development. In the abstract, these targets are blunt instruments; they need design interventions to engage the imagination and empower action.

From restoration sites to rewilding initiatives, from greenways to green infrastructure, we must engage in nothing less than a planetary strategy of landscape connectivity. Large wilderness is now rare, but its interstitial spaces will be the practice of the everyday. Designing and remaking connections between remnant wild fragments will be paramount—from the mongrel places[1] of the in-between, to novel and hybrid ecosystems, to agricultural working lands, to reserves for hunting and harvesting, and even to derelict places of urban decay. Together these landscapes will form a wild mosaic for the next wave of conservation. In the Anthropocene, there is no *away* to which we can retreat, no pristine place unaffected by human hands. Rather, we need design tactics for the full spectrum of landscapes from urban to suburban to rural to wilderness. The old wilderness is now but fragments, and the wild (and its qualities) will be found in the refuges and connective tissue in between. The local work of the landscape architect will be humble, to stitch together the fragments, but the cumulative design is planetary: we must (re)weave the tapestry of the wild back into the landscape of the future.

To lose the wild is to lose that which makes us most human. The sad irony is that in wasting the wild, we lose a vital, visceral, and primal part of ourselves. Landscape architecture has the tools to integrate these stories through the medium of design, reflecting the relationship between wild places and the emotional responses they provoke—and the very human qualities they evoke. Reflected in art, anchored in master plans and policies, implemented in design, landscape architecture has the power and authority to make the story of the wild legible, to re-center its place within the landscapes we make, and, by extension, to wake the wild within the human.

So I urge us, as landscape architects and allies, to reaffirm the primordial place of the wild, reactivate the vital role of wild things, and reconnect the landscapes that sustain us all. In so doing, we must design with awareness, humility, intention, direction, and conviction. To honor the voice of the wild, we must listen for it. To reveal the sublime of wild places, we must see them. And to assert the *wild-ness* that makes us human, we must value it. For without the wild, we are condemned to the endless monochrome, lost to a monoculture of our making.

> To lose the wild is to lose that which makes us most human. The sad irony is that in wasting the wild, we lose a vital, visceral, and primal part of ourselves.

1.   Richard Weller's term, elaborated in Richard Weller, "World P-Ark," *LA+* Wild 1, no. 1 (2015): 10-19.

Nina-Marie Lister is graduate program director and associate professor in the School of Urban and Regional Planning at Ryerson University in Toronto. She is a registered professional planner trained in ecology, environmental science, and landscape planning, and the founding principal of PLANDFORM, a creative studio practice exploring the relationship among landscape, ecology, and urbanism.

Chapter 5

THE ENVIRONMENTAL/SOCIAL CRISIS AND THE CHALLENGES OF INFORMAL URBANIZATION

DAVID GOUVERNEUR

If we want to have a better life, our neighbors must live better also.
—Oscar Grauer

The disparities between the haves and have-nots are growing in technically advanced nations, while the gap separating these countries from the developing world is widening.

Politically unstable, war-torn areas of the Global South often lose people migrating to the developed world. Urban inequality results in increased levels of social resentment and violence. These are perhaps just early indicators of what the future may bring. Such conditions will worsen as a consequence of climate change, water and food shortages, and the growing disparities in access to infrastructure, services, new technologies, information, and appropriate governance.

This bleak forecast and the depletion of the planet's resources are closely tied to urbanization at exponential levels, much of which will take place in developing countries where some of largest urban agglomerations in history already exist. A high percentage of this growth will occur in the form of informal settlements.

In the early twentieth century, modern city planning was globally recognized as a tool to manage rapid urbanization. Planning emphasized the functional and quantitative, ignoring cultural and environmental nuances. Zoning was introduced to regulate land uses, correlated with mobility and infrastructure systems and services. An unintended consequence was

After decades of informal settlements being ignored... informality became the dominant form of urbanization, and creative planning and design approaches emerged to improve existing informal areas.

the expansion and fragmentation of once compact cities. Additionally, it became a tool for social segregation.

In the developing world, where a high percentage of the population could not participate in the real-estate-driven model, planning unintentionally pushed out the poor from the areas designated for urban expansion. Settlements occupied peripheral sites unfit for urbanization: unstable land, floodplains, and areas next to landfills, under high-power lines, or in environmentally sensitive and protected areas.

While the informal city evolved with great dynamism, presenting strong social ties, encompassing informal economies, and producing compelling organic forms, it did so

without the basic advantages of urban living. Informal cities have limited accessibility and mobility, endure unsanitary conditions, and lack public space and services, resulting in urban areas that are incomplete and submissive to the formal city. As informal areas become larger, inequality increases and city performance falls behind.

After decades of informal settlements being ignored, eradicated (only to reappear someplace else), and targeted by housing programs with limited coverage, informality became the dominant form of urbanization, and creative planning and design approaches emerged to improve existing informal areas. Latin America was at the vanguard of such trends. In 2014, for example, the World Urban Forum was held in the one-time murder capital of the world—Medellín, Colombia— which proudly demonstrated how their interventions reduced levels of violence and social inequities and improved overall city performance.

These trends are now being emulated in many developing countries; however, these successful urban interventions are laborious and have limitations. Such changes are efficient at a neighborhood scale, but they cannot address larger, more complex urban areas. In addition, an estimated one billion people will live in new informal settlements over the next two decades. Even in countries where the government proactively deals with informality, officials are skeptical of planning ahead for new informal areas. Officials are unable to deter the informal areas and allow them to spring up, thinking that perhaps in the future they can address resulting problems with settlement improvement projects.

A paradigm shift is urgently required. In *Planning and Design for Future Informal Settlements: Shaping the Self-Constructed City* (Gouverneur 2014), I suggest how such challenges may be handled. The proposal is simple but entails a different mindset as well as new design and managerial criteria, which I have referred to as the informal armatures approach (IAA).

IAA suggests that emerging informal areas, properly assisted, can evolve into balanced and attractive urban areas, perhaps surpassing the performance of formal cities. It is possible to combine the vibrancy, flexibility, organic morphology, sense of place, and communal engagement of informality with the benefits of cutting edge planning and design. This preemptive and ongoing method may nurture a hybrid product (formal/informal), operating at different scales (from the site specific to the territorial), and resulting in a rich urban ecology, in constant transformation, while increasing resiliency.

For IAA to succeed, certain conditions should be met. There must be acceptance that inaction will result in a high price for social and environmental stress, which requires acknowledging that informality is a valid form of city making and that we must embrace it with creative ways of influencing it. Proactive land-banking policies must be employed to reduce the exclusionary effects of the real-estate-driven model to attain, over time, more balanced urban organizations. Planning, design, and managerial efforts must focus only on what communities cannot address alone, with a physical and nonphysical system that will support rapid change by addressing the public realm, the delivery of services, and how these transform over time; providing patches where self-construction is suitable; and gradually incorporating urban components that usually only exist in the formal city.

> The informal armatures approach suggests that emerging informal areas, properly assisted, can evolve into balanced and attractive urban areas, perhaps surpassing the performance of formal cities.

And alliances must be formed among the political sector, the professional milieu, the private sector, and communities, with strong leadership and qualified facilitators who can work on-site with cross-disciplinary teams and residents.

In contrast to conventional planning, this approach enhances the environmental, the social, and the performance features, delving into aspects that are relevant to each context, envisioning a compelling, flexible, and transformative public realm.

These supporting armatures may bundle different components: low-cost mobility, water management, food production, access to education, local manufacturing skills, reduction of violence, enhanced self-governance, etc. For the most part, communities will do what they know best: they will construct their own homes, develop strong social ties, and incorporate income generating activities.

IAA can be considered a landscape urbanism method that can better assist the emerging informal city. While these ideas are particularly helpful to the developing world, the basic criteria—addressing efficacy, creating added value, managing constant transformations, and fostering resiliency—may be applicable in any context.

David Gouverneur is associate professor of practice in the Department of Landscape Architecture at the University of Pennsylvania School of Design. He is also professor emeritus of Universidad Rafael Urdaneta in Maracaibo, Venezuela. Gouverneur received his architecture degree from Universidad Simón Bolívar, Caracas, and his master of architecture in urban design from Harvard Graduate School of Design.

Chapter 6

EVOLUTION AND PROSPECTIVE OUTLOOK

MARIO SCHJETNAN

The Landscape Architecture Foundation's *1966 Declaration of Concern* established very clearly the concern for poor environmental conditions, social inequalities, and loss of quality of life prevalent in most North American cities at the time. It was a timely and valorous call, a moral outcry by landscape architecture leaders of the time.

To be sure, many US cities in the past 50 years have improved their levels of air quality, decreased their contamination of soils and water, and upgraded their public open spaces. Many cities have rehabilitated and repopulated their city centers and become more habitable in general. However, many other challenges and global concerns have now arisen, including climate change and the horizontal expansion of cities; and the United States still maintains the highest levels of consumption in the world, per person, of natural resources, energy, land, and water.

Fifty years ago, Latin America had few landscape architects and not a single organized professional society or specialized school. Today, there are several universities offering master's degrees in landscape architecture and 16 associations registered with the International Federation of Landscape Architects (IFLA). However, in the past 50 years, the damage to the environment has grown exponentially, with desertification and the loss of many ecosystems, including enormous chunks of the Amazon jungle, rainforests, wetlands, and mangrove swamps. Deterioration affects not only local and regional communities but also environmental conditions at the global level. (In Mexico, 300,000 to 400,000 people abandon their land each year because of soil degradation—erosion and salinization.)

Our...declaration must be global because much of the environmental impact and urban growth is happening in the developing world.

The urban environment in Latin America has seen phenomenal expansion. Today, Latin America has sixty-six metropolises of more than one million inhabitants and five megalopolises with ten million inhabitants or more (Buenos Aires, Lima, Mexico City, Rio de Janeiro, and São Paulo). Although many landscape architects actively participate in government and private or social organizations, their numbers are small—disproportionately small compared to the amount of environmental problems and conditions of large numbers of urban dwellers. This is unacceptable. While there are 150,000 architects in Mexico alone, fewer than 1,000 landscape architects work in the country. (In 2015, the Urban Mobility Observatory of the Development Bank of Latin America studied 15 metropolises and found that these cities had problems of congestion and contamination, with 24 million cars, 1 million buses, and 590,000 taxis. The inhabitants lost more than 118 million hours a year just in transportation.)

Most of the global urban expansion in the next 50 years will be in the so-called developing economies: Africa, China, India, and Latin America. According to the *Economist*, developing economies already surpass developed economies or the so-called rich countries—including the United States, European countries, Australia, and New Zealand—in multiple indicators such as consumption, exports, imports, oil, steel (75 percent), and cement. In 2011, subscriptions to mobile phone services topped 82 percent, and 52 percent of all purchases of motor vehicles occurred in these economies.

Therefore, our call or declaration must be global because much of the environmental impact and urban growth is happening in the developing world. Eurocentric or US-centric visions must change if landscape architects aspire to influence in meaningful ways the development of urban places and the viability and conservation of world resources. Among the issues to consider are that green technologies, sustainable societies and cities, and the green economy are terribly relevant for our means and aspirations; a green economy and technology

Landscape architects must be educated, enabled, and fit to lead an authentic and urgent green evolution for a sustainable, viable, and just planet.

can provide jobs and generate new economies and synergies. (In 2015, Mexico exported more dollars in the agroindustry than in the petroleum or tourist industries.) In addition, the profession of landscape architecture is a major player, a means to achieve environmental justice and social equality, as well as avoid urban and rural marginality. Landscape architects must be educated, enabled, and fit to lead an authentic and urgent green evolution for a sustainable, viable, and just planet; and substantial numbers of candidates for degrees in landscape architecture will need to study in North American universities and return to their respective countries.

The Landscape Architecture Foundation can document and disseminate cases of successful interventions where landscape architects have improved and positively transformed communities, environments, landscapes, portions of cities, as well as people's lives, particularly in developing countries.

Mario Schjetnan is founder and director of Grupo de Diseño Urbano/GDU, an interdisciplinary design group connecting landscape architecture, architecture, and urban design in spatial, aesthetic, and social contexts. He earned a degree in architecture from the National Autonomous University of Mexico (UNAM) and a master of landscape architecture, with an emphasis in urban design, at the University of California, Berkeley.

Chapter 7

THE LARGE-SCALE CENTURY AHEAD

ALAN M. BERGER

Landscape architects face an urgent choice. Will we create the large-scale solutions to the ecological challenges of the coming years? Or will we abdicate that role to engineers, scientists, and technocrats? Despite 20 years of progress in ecological design, we are now in danger of limiting our interventions to small-scale achievements in the historical tradition of art and form making. Instead, we could follow another thread in our profession's history, making contributions on the sort of scale needed to meet the vast environmental challenges of the next century.

To make this change, landscape architecture must renew its commitment to bioregional research and practice. Climate change and global urban expansion mean a revolution spurred by an oncoming tide of water and the necessity of relocating whole populations. Who will define the strategies necessary to accommodate these changes? Landscape architecture has the opportunity to provide large-scale answers if we do not step back fearfully in the name of artistic—or egotistical—imperatives.

We live in an increasingly altered natural world: a new geologic era called the Anthropocene, with the entire earth experiencing accelerated environmental change. Frequent catastrophic events coupled with a new global willingness to act collectively make large-scale initiatives achievable—as they were in the past—for landscape architects.

If this is to be accomplished, landscape architecture's relationship to the science, technology, engineering, and mathematics (STEM) fields must be reinvigorated. We no longer teach scientific and quantitative methods in design schools, a tragedy unimagined by our predecessors.

Designing with the intent of creating only a general awareness or an aesthetic appreciation of natural systems is not a strategy for future success. We have already seen steep declines in how children engage with the natural world over the past decade. Reopening our field to scientific thinking

If landscape architecture is to have any agency in defining and creatively solving large-scale environmental problems, it will need a sound foundation of scientific reasoning and quantitative research as part of its design culture.

will broaden its appeal to a new generation, one that is environmentally motivated. If landscape architecture is to have any agency in defining and creatively solving large-scale environmental problems, it will need a sound foundation of scientific reasoning and quantitative research as part of its design culture. Otherwise, we risk allowing the engineering fields full agency over solutions.

Science must be the base for a new competence in designing for large-scale climate impacts. Astonishingly, J. B. Jackson spoke about climate change to an audience of landscape architects almost exactly 50 years ago. He wisely observed that "the beaches along the Gulf and the Atlantic... are being remorselessly eaten away by the sea. Geographers tell us that we are being driven inward by the rising ocean at a rate of one vertical foot a century...." Jackson urged the field to embrace change by designing for it, rather than trying to preserve the landscapes of the day.[1] Yet, landscape architects

around the world today are designing new waterfront parks filled with old paradigms of native vegetation that dies and landforms that erode when the first salty storm surge hits, even as scientists conservatively estimate a sea-level rise of ten vertical feet by 2065.

New lines of large-scale practice should be built around resilience adaptation, reminiscent of Frederick Law Olmsted, Jr.'s ambitions, Christopher Tunnard's regionalism, Garrett Eckbo's primeval expansion, Charles Elliot's infrastructures, and Ian McHarg's ecosystems. We have excellent precedents for a type of landscape architecture that is more active in large-scale design. We need to create a real value, a quantitative difference, that decision makers will find worth buying. Good design can certainly do this, too, but High Line commissions are few and far between. If the field is to recover its large-scale influence, it must once again open collaborative lines of engagement with urban planners who deal directly with economics and policy. I can hear the collective groan. Tunnard foresaw a similar audience reaction to large-scale agency while speaking at Harvard in 1965.

Nonetheless, we should seize the opportunity presented by climate change crises. Populations will be forced to transfer, and cities will need to be retrofitted for massive urban flooding. Fifteen of the world's twenty megacities are located in coastal zones. In the United States, 23 of the 25 most densely populated counties are coastal. *Resilience* is finally emerging as a new field of research and practice, and landscape architecture and urban planning must lead together, with the backing of the sciences.

Today's digital world has also radically changed the way we see and measure natural and human processes. Environmental problems grow ever larger in scale, and technology allows us to interpret changes and intervene creatively. Yet, in order to fully embrace technology and big data relative to the landscape medium, we need new collaborators in computer science and in technology. In my current position at the Massachusetts Institute of Technology, I see technology unfolding in new ways every day—in the personal cube satellites, the sensing capabilities, drone development, autonomous and artificial intelligence, and the ubiquitous Internet of Things. These innovations create a new connectivity of objects across landscape space that flows through our daily experiences. All of us, particularly future teachers, must become fluent technology innovators and accelerate new practices.

During the next fifty years, an estimated three billion more people will live in urban areas across Africa, India, and Latin America. Is landscape architecture going to lay the ecological and infrastructural foundation for the next great wave of urban expansion? Governments will have to spend about $71 trillion by 2030 to provide adequate global infrastructure for electricity, road and rail transport, telecommunications, and water. If our field is to expand intelligently, it must invent original, empathetic models for the developing world that take into account the abundant bioregional resources before they are destroyed. Perhaps large-scale landscape practice can be better achieved in developing contexts, with models of sustainable development integrated with new urbanization.

Designing with the intent of creating only a general awareness or an aesthetic appreciation of natural systems is not a strategy for future success.

When I travel in those regions, I ask folks if they know what landscape architecture is. Other than as gardening, they are unfamiliar with the profession. My challenge to the field is to expand landscape architecture both within the United States and to these continents while avoiding the design hegemony that took place in China over the past decade. Let us renew our commitment to large-scale visionary work, and let us do it right.

1.   J. B. Jackson, "Can The Natural Environment Be Saved?," in *1965 Urban Design Conference* (Harvard Graduate School of Design: 1965), 68.

Alan M. Berger is Norman B. and Muriel Leventhal Professor
of Advanced Urbanism in the Department of Urban Studies
and Planning at the Massachusetts Institute of Technology.
He is codirector of Norman B. Leventhal Center for Advanced
Urbanism (LCAU) and founding director of P-REX lab at MIT.

Chapter 8

DEVELOPING LANDSCAPES OF RESOURCE MANAGEMENT

ALPA NAWRE

With what are we welcoming our future generations? Piles of plastic? Polluted air and dirty water? Life in degraded environments with mismanaged resources is the normal human experience in many parts of the world. The statistics are staggering. Of the total world population of 7.2 billion, about 6 billion people live in developing countries where access to clean water, clean air, and efficient systems of waste disposal is a daily struggle.

Water, especially, is a severely contested resource in these contexts, both in terms of quantity and quality. In India, for example, over 100 million people lack access to safe water, and diarrhea causes 1,600 deaths daily. Where water mafia and dacoits are a grim reality, where suicides, murders, and street fights over water scarcity are a serious issue, and where commuting to and from work could involve wading through chest- or knee-high floodwater, the problems associated with water management in India point to a crisis that is only expected to get worse with impending climate change impacts and rapid urbanization. And while some problems clearly fall outside the scope of a landscape architect, there are many issues that can be addressed through better water management landscapes. This is where the agency and action of landscape architects at both the systems and the site scales become critical, applicable not only to water but also to other contested resources.

Today, in developed countries we are shocked by and somewhat resigned to reports and personal experiences of Beijing's air quality, the water crisis in India, or food scarcity in Africa. Conditions, however, were not so very different in the 1950s and 1960s in North America when people wore gas masks in Los Angeles and decried the region's filthy rivers. When a small group of landscape architects gathered in Philadelphia and crafted the *1966 Declaration of Concern*, noting the degradation of America's water and air, the world was not such a different place. Now, however, the issues have become more global, critical, and widespread.

In this context of contested resources, landscape architects must step in to do what we can to restore and reestablish healthy relationships between humans and their environment. I entreat all landscape architects to rise above parochial discussions, territorial predispositions, and disciplinary comfort zones to address the very real issues of water, air, food, waste, minerals, and energy, with which rapidly urbanizing and developing countries such as India now grapple.

The *1966 Declaration of Concern* is a demonstration of the enormous responsibilities that the profession attempted to take on. The last 50 years have seen the coming of age of the landscape architecture profession. We have drawn on formidable skills of research and analysis to understand and map multilayered issues, and we have conveyed this understanding to the general public through visualization of complex landscape systems spanning both scale and time. Many landscape architects have attempted to restore damaged ecosystems and design better human/nonhuman habitats. Yet, we have just scratched the surface, and much remains to be done in the context of resource management, especially that of water, food, and waste in developing countries.

From these countries, there are many lessons to be learned on alternative definitions, frames, paradigms, systems, and landscapes of resource management, all of which are rapidly being transformed and degraded as we speak. We urgently need to understand the various ecologies of resource management in the developing world. What can we learn from cultures that designed multifunctional resource infrastructure and practiced community ownership of landscapes to inform the design of resource management in industrially developed

> In the context of contested resources, landscape architects must step in to do what we can to restore and re-establish healthy relationships between humans and their environment.

countries and vice versa? Before we engage in design, we must understand and evaluate existing systems.

As designers, we have two avenues of intervention for addressing resource issues. The first is through design to improve existing resource landscapes, and the second is to create alternative paradigms for better resource management through the structuring of new built environments. The projected increase of the world's population to nine billion by 2050 will be almost entirely from developing countries, accompanied by rapid urbanization. For example, in the next 50 years, India's population will peak at 1.6 billion and the country will be adding more than 400 million to its urban population—about 20 more Mumbais! The development of urban territories to accommodate these millions desperately needs the expertise of landscape architects equipped to design urban landscape systems for better resource management. It also presents unprecedented opportunities for design experimentation. How do we take the lessons we have learned in the urbanization of developed economies and apply them in our design responses to the resource management problems of the developing world?

Part of the challenge is not only to address resource management issues head-on but also to make the public—especially the decision makers in the developing world—aware of the contribution that we can make in improving resource management.

In most parts of India, when I introduce myself as a landscape architect, people either transform the phrase to *landscaping* or *gardening* or latch on to the familiar *architecture*—not surprising since there are very few landscape architects in India. About 800 landscape architects serve a total population of 1.25 billion, and of this handful, fewer still engage in issues of resource scarcity and/or mismanagement. As landscape architects, we must actively make opportunities

> Today's landscape architecture students live in a complex, networked world and must be prepared for a future defined by global professional practice, to meaningfully engage in and to craft the built environment of not only their own community but also of cultures dramatically different from their own...

for engagement happen by better preparing ourselves with alternative design solutions and communicating them to the public.

Today's landscape architecture students live in a complex, networked world and must be prepared for a future defined by global professional practice, to meaningfully engage in and to craft the built environment of not only their own community but also of cultures dramatically different from their own—dealing with life-threatening issues related to water, food, and waste. These issues often fall outside a landscape architect's traditional scope (a missed opportunity for the discipline). Training the future generation of landscape architects to deal with these issues at different scales is the only way to make our discipline relevant in the coming 50 years.

It is an exciting time to be a landscape architect, but only if we embrace the opportunities and challenges ahead of us. There must be a crusading determination on the part of landscape architects to address the real issues of resource management if we are ever to permanently establish and realize the true potential of our discipline.

Alpa Nawre is assistant professor in the Department of Landscape Architecture and Regional and Community Planning at Kansas State University and partner at her design practice, Alpa Nawre Design. Nawre holds a master of urban design from Harvard Graduate School of Design, a master of landscape architecture from Louisiana State University, and a bachelor of architecture from NIT, Raipur, India.

Chapter 9

LANDSCAPE ARCHITECTURE AS NECESSITY

KELLY SHANNON

The authors of the Landscape Architecture Foundation's *1966 Declaration of Concern* claimed they were brought together by a highly disturbing environmental crisis. They issued a call to arms for the profession of landscape architecture. Fifty years later, in a world that is increasingly divided, landscape architecture arguably remains the most powerful tool to marry social and ecological justice.

Landscape architecture has the capacity to address the most pressing and fundamental problems facing the world today: climate change, water and food security, afforestation, infrastructure, energy, and unchecked urbanization. An entirely new array of challenging programs and projects await the landscape architect. Moreover, in addition to designing for beauty and function, there is persuasive political agency of well-designed territories that work intelligently across scales and ecologies.

Landscape architecture involves systemic thinking, concerned with ecologies of dynamic change, with adaptability, resilience, and flexibility, while at the same time giving the necessary attention to the concrete and its materiality. Today we live in remarkably turbulent and critical times that require radical system change: politically, socioculturally, economically, and spatially. The status quo is unsustainable. For humanity. For planet Earth. For future generations of both.

We are now at a precipice, a critical tipping point, standing before the wholesale destruction of the planet. The never-ending, interlinked, and adaptive cycles of growth, accumulation, restructuring, and renewal occur in human and natural ecologies of nested sets of scales and timeframes.

Landscape architecture must become *the* essential game changer in not only reshaping the earth's ecological systems in practice but also in transforming the fundamental habitation of the planet through broader systemic thinking.

The world is marching toward catastrophe, and landscape architecture has a responsibility to be at the core of a categorical shift. Climate change is not pseudoscience. Compromised summit declarations do not have a forceful

> Landscape architecture must become *the* essential game changer in not only reshaping the earth's ecological systems in practice but also in transforming the fundamental habitation of the planet through broader systemic thinking.

impact. Landscape architecture must set the boundaries for policy makers and orient social movements. Bold and inspired projects can lead policy.

The public context in the Global North requires careful yet bold reediting and recalibrating because the process of development transforms, optimizes, and corrects; the process is not primarily concerned with accommodating massive demographic, social, and economic growth in new urban areas. New hybrid ecologies must create innovative synergies between city and nature, between consumptive and productive space, between impermeable and porous surfaces, and between the urban and the rural.

In the Global South, where urbanization, industrialization, and migration gallop ahead, stabilizing natural ecologies must take place simultaneously with massive development. Robust green and blue systems can structure territories and create frames for adaptive landscapes. Resilience must be built into territories and cities by designing an overlap of natural and built ecologies.

Despite the impassioned plea of our colleagues 50 years ago, landscape architects have mostly been peripheral to large environmental agendas; in Western Europe, the field is at its most progressive with landscape architects occasionally leading major urban transformation projects and participating in robust infrastructure projects. The American landscape architect must more proactively create larger opportunities for the design of the environment beyond the piecemeal and ad hoc, and landscape architects must deepen their commitment and the political will for the continued expansion of the public realm. Clients must be cultivated, and the role of the landscape architect must be properly understood and valued at the national, regional, city, and local levels.

Immense opportunity and obligation await the American landscape architect in particular. Increasing civic discord across the nation's cities and the effects of recent climate-related extremes like prolonged heat waves, droughts, floods, cyclones, and wildfires reveal the significant vulnerability and exposure of human and natural ecosystems. Taken together, this is a renewed wake-up call, and landscape architecture is urgently needed to create equitable and truly democratic public realms and to implement risk mitigation strategies.

At the same time, it is landscape architecture that has the ability, and, indeed the responsibility, to ground projects to their specificity: their workings, secrets, logic, opportunities, and inherent dangers. The "green wave" and corresponding performance-driven obsessions must not be a panacea. Fieldwork unveils specific geologies and geographies of contexts and the messiness and contested realities of the everyday; investigations into the existing logic of landscapes, cultural appropriations of a territory, social formation and codification of spaces require both diachronic and synchronic perspectives.

On-site interpretative mapping can go far beyond the descriptive and become a form of critical realism (critical in selecting what to map) and the base for insights through the discovery of unspoken and unwritten realities gained from a haptic and experienced sense. In turn, landscape architecture is the operative instrument for engaged resistance to relentless homogenization, globalization, and "flattening out" of cultures and places. Projects must be contextually embedded geographically, geologically, and culturally.

Landscape architecture is an ever more urgent necessity. It is simultaneously necessary for landscape architecture to reclaim its position as a key discipline—able to uniquely synthesize ecological systems, scientific data, engineering methods, social practices, and cultural values—and integrate them all into the design of the built environment. Bold visions must combine the tangible and the imaginary to provoke conversations that promote social equity and environmental justice as well as manifest landscape's transformative power. It is time once again to make a strong call to arms to the profession to more actively engage with the world and prove that landscape architecture is a necessity. The political agency

The political agency of the profession must be forcibly reactivated and the power of landscape architecture engaged to be *the* game changer in reshaping ecological systems and transforming forms of living.

of the profession must be forcibly reactivated and the power of landscape architecture engaged to be *the* game changer in reshaping ecological systems and transforming forms of living.

Kelly Shannon, PhD, is professor of architecture and director of the Graduate Program of Landscape Architecture and Urbanism at the University of Southern California. She earned her architecture degree at Carnegie Mellon University, a post-graduate degree at the Berlage Institute in Amsterdam, and a PhD at the University of Leuven, where she focused on landscape to guide urbanization in Vietnam.

Chapter 10

THE ART OF SURVIVAL

KONGJIAN YU

The past 50 years have been fruitful for the profession of landscape architecture. Compared to the environmental situation that our ecological planning pioneers faced 50 years ago, North American and European countries, where urbanization and industrialization had led other parts of the world, have seen great improvement. The profession of landscape architecture has been key to these improvements by raising public awareness and improving the practice of planning and designing our natural and human ecosystems. In addition, we have learned to manage our natural and cultural assets better using a variety of techniques, such as design with nature, sustainable site management, ecological stormwater management, greenways and green infrastructure, green roofs, and community engagement. Such obvious success has proved what had been projected by our

forerunners 50 years ago in the *1966 Declaration of Concern*: "There is no one-shot cure, nor single-purpose panacea, but the need for collaborative solutions."

But the challenge to survival is not yet over. The improvements at the limited regional scale in the United States cannot remediate the deteriorating global situation. Today, along with globalization and worldwide urbanization, what had been regional and national environmental issues 50 years ago are now global and are particularly troublesome in developing areas such as China, India, and Southeast Asia. Moreover, the severity of the changes, such as increasing water, food, and energy shortages, coupled with environmental degradation, threaten the survival of humanity itself. In China, for example, within the past 50 years, the population has doubled, the population in urban areas has increased six-fold, 75 percent of the surface water is polluted, and 400 cities suffer water shortages. At the same time, 70 percent of the population struggles with urban and rural flooding every year, one-third of the nation is under threat of heavy smog, 50 percent of wetlands have disappeared in the past 50 years, and a significant number of rare species will never be seen again. In addition, China has suffered a significant loss of much of its rich cultural heritage. The leaders of the profession said half a century ago that "what is merely offensive or disturbing today threatens life itself tomorrow," and it is clear that we have no place to escape.

We are, however, not helpless. What our pioneers have done in North America since the 1966 declaration proves to the world that the charge launched by the profession of landscape architecture against environmental degradation is successful, and an integrative and symbiotic approach, namely, the design of landscape processes and patterns, is the key. What we need to sustain us in the next charge against global environmental degradation is to replicate what has proved successful in best practices, opening up ever more integrative and symbiotic approaches at even larger scales and working even more comprehensively. More than at any other time in history, we must fulfill the mission of healing the earth, a mission defined by our forefathers, where the landscape architect should be a "conductor," as Ian McHarg declared, not a solo artist, who can

bring all related disciplines and individuals into the medium of landscape.

To face such global challenges and opportunities, we have come to the point of redefining the profession of landscape architecture as the art of survival: to heal the earth and sustain humanity. Landscape is the medium where all natural, biological, and cultural processes interact, and landscape architecture (the planning, design, and management of the landscape) is, therefore, the profession that could take the lead

> What our pioneers have done in North America since the 1966 declaration proves to the world that the charge launched by the profession of landscape architecture against environmental degradation is successful, and an integrative and symbiotic approach, namely, the design of landscape processes and patterns, is the key.

in meeting the challenges of survival in a complicated physical and cultural environment.

In facing contemporary ecological and environmental issues, three strategies will allow our profession to take leadership. First is planning practice. We must secure and consolidate across scales an integrated ecological infrastructure that safeguards various ecological and cultural assets and provides effective multiple ecosystem services, which will be the tool and framework for smart growth and smart preservation.

Second, design and management practice. We need to define a new aesthetic, the "big feet" aesthetic (the aesthetics

We have come to the point of redefining the profession of landscape architecture as the art of survival: to heal the earth and sustain humanity.

various climate and environment situations, and let them build a database to share among professionals the best practices for facing various environmental challenges. These will open up a wholly new horizon for the profession of landscape architecture.

of sustainability), as opposed to the distorted "little feet" aesthetics of ornamental, calling for deep forms (as opposed to the costly shallow or fake forms) with human desires based on healthy ecological processes; this aesthetic must be high performance and productive, able to provide critical ecosystems services with little maintenance.

And third, global cooperation and education. As professionals and academics, we must cooperate at the global scale, sharing knowledge and working together in a real battle against environmental challenges at any corner of this globe. We know that any change in one part of the globe will affect us in another part; we must reform and restructure the education programs of landscape architecture to build and consolidate the bedrock of landscape architecture as an art of survival, through building and sharing the wisdom of survival skills and techniques in dealing with floods, drought, and food production, instead of teaching the ornamental art of gardening and pleasure-making of an elite culture. Let peasantry and sustainable vernacular landscape practices show students the knowledge and windows for survival in

Kongjian Yu, DDes, is founder and dean of Peking University College of Architecture and Landscape. He founded Turenscape, the first private firm of its kind in China, and practices globally. He received his doctor of design from Harvard Graduate School of Design.

PART III

THE LANDSCAPE

SIMULTANEOUSLY, THERE IS PROFOUND HOPE FOR THE FUTURE. AS WE BEGIN TO UNDERSTAND THE TRUE COMPLEXITY AND HOLISTIC NATURE OF THE EARTH SYSTEM AND AS WE BEGIN TO APPRECIATE HUMANITY'S ROLE AS INTEGRAL TO ITS STABILITY AND PRODUCTIVITY, **WE CAN BUILD A NEW IDENTITY FOR SOCIETY AS A CONSTRUCTIVE PART OF NATURE**.

THE **URGENT CHALLENGE BEFORE US IS TO REDESIGN OUR COMMUNITIES** IN THE CONTEXT OF THEIR BIOREGIONAL LANDSCAPES ENABLING THEM TO ADAPT TO CLIMATE CHANGE AND MITIGATE ITS ROOT CAUSES.

FROM THE NEW LANDSCAPE DECLARATION

Chapter 11

BOUNDARY EXPANSION

CHARLES A. BIRNBAUM

Important questions about the cultural value of our shared landscape heritage are finally beginning to move from the province of theoretical arguments in the pages of scholarly journals to discussions in city councils, on editorial pages, in blogs, and elsewhere in the public realm. Increasingly and incrementally, the cultural value of landscape architecture is being reevaluated as the resurrection of the nation's urban centers continues. What role should landscape architects and the academic and educational organizations that serve them play going forward?

As we reveal and explore the fundamental issue of cultural value—an issue that has received less attention than natural resource value, or what Ian McHarg referred to as "the ecological view" in *Design with Nature*—we must begin by recognizing that our ability to adequately assess and quantify the cultural value of landscape architecture is a work in progress. If we aspire to meet McHarg's "quest for survival, success, and fulfillment," I suggest that a more holistic, systems-based approach to problem-solving and planning is essential. This is an opportunity to fundamentally expand upon the *1966 Declaration of Concern* and McHarg's lofty vision.

In 1968, Grady Clay wrote the following in his editorial "Who Says 'Never Look Back'": "To be 'forward looking' in our society is to march with its Establishment—even to destruction or infamy. To look backward in anger or in spirit of inquiry is discouraged by generations unconcerned with yesterday. Yet it should be clear that yesterday is crammed with environmental miscalculations and ascertainable errors, as well as successes. We would do well to admit that this country needs more formal, effective, and organized systems for reviewing and evaluating physical development and change. We need them so that we may improve our record of lousing up the landscape repeatedly, using the same tactics, making the same mistakes generation after generation with a familiar mixture of cupidity and stupidity."[1]

The time has come for us to expand beyond our understanding of just ecological systems to include the evaluation, quantification, and assessment of the historic and cultural value of landscape architecture.

Let us step back and consider the cultural context of the declaration itself: the year 1966 gave us the movie *Born Free*, the song "California Dreamin'," and *Don't Drink the Water*—a celebrated Broadway farce by Woody Allen. Meanwhile, within the profession, Lawrence Halprin published *Freeways* and unveiled Lovejoy Plaza in Portland, Oregon, while, back east, Jacob Riis Plaza by M. Paul Friedberg opened in New York City as did Boston's Copley Square by Sasaki, Dawson, DeMay and Associates.[2]

Exciting opportunities were in the air, reflected in ambitious projects of diverse types, scope, and scale—from the ecologically-driven Sea Ranch in California and The Woodlands in Texas to the reimagining of the public realm in our cities with Paley Park (New York City), Nicollet Mall (Minneapolis), Golden Gateway/Sydney Walton Park (San Francisco), and the Art Institute of Chicago South Garden.

Another significant development, however, completely absent from the 1966 declaration, was the sweeping National Historic Preservation Act of that same year. The enabling legislation recognized that "the spirit and direction of the nation are founded upon and reflected in its historic past" and that "the historical and cultural foundations of the nation should be preserved as a living part of our community life and development in order to give a sense of orientation to the American people."[3] This consequential law, which emphasized the value of the nation's "historical and cultural foundations," marked the inauguration of the National Historic Landmarks program, with designations for Niagara Reservation and Central Park. In addition, an Olmsted revival started to take hold, spurred by *Man and Nature: The Olmsted Exhibition* organized at Harvard Graduate School of Design and curated by Professor Albert Fein, with assistance from the ASLA.

Unfortunately, despite the growing reverence for Olmsted, Fein determined that the same could not be said of our shared landscape legacy. In an October 1972 article[4] in *Landscape Architecture Magazine* about a Ford Foundation/ASLA Foundation-funded "Study of the Profession" (begun in January 1969), Fein lamented "the discovery of the view of the past held by most landscape architects. According to the Gallup Survey, it is almost uniformly viewed as being of least importance in the training of a professional...it has not yet been accepted by this profession that it is part of a historic stream; that history is everything that happened up until a minute ago; and that in this case, therefore, includes the work of such contemporaries as Halprin, McHarg, Eckbo, Sasaki, Simonds, Church, and others—that a denial of history is a denial of the civilized mandate to constantly re-examine what we have done in terms of what we are and wish to become. And this failure of reappraisal has contributed on numerous levels to the deterioration of quality in public land design. Without an adequate appreciation of history, it is difficult to understand how members of any profession can view themselves as part of a humane and scientific discipline."

As we mark the declaration's half-century anniversary and the impending one for McHarg's seminal book *Design with Nature*,[5] the time has come for us to expand beyond our understanding of just ecological systems to include the evaluation, quantification, and assessment of the historic and cultural value of landscape architecture. Going forward, universities and professional offices should reject antediluvian divides, adopt a more holistic mindset, and place a value on our landscape heritage.

"By raising the level of awareness of the past," Fein wisely observed, "landscape architecture will be making a substantial contribution to a more popular understanding of its present and future goals." History can, he continued, "enable landscape

architecture to better understand the points at which it has interacted—or failed to interact—with other organizations and/or institutions that should complement the goals and functions of the profession."[6]

Declaration cosigner Grady Clay noted: "A principal recommendation of the [aforementioned 1969] study [of the

> Going forward, universities and professional offices should reject antediluvian divides, adopt a more holistic mindset, and place a value on our landscape heritage.

profession] is that landscape architecture be redefined as the art of applying scientific principles to the land—its planning, designing, and management—for the public health and welfare, with a commitment to the concept of stewardship of the land."[7]

The time has come for *stewardship* to embrace and incorporate natural and cultural resource values in our quest for survival, success, and fulfillment.

1. Grady Clay was one of the original authors of the *1966 Declaration of Concern*. He published this editorial in the January 1968 issue of *Landscape Architecture Magazine*.

2. Re-evaluating the Jacob Riis Houses in a 1999 editorial for *Land Forum*, Peter Walker proclaimed that, for landscape architecture, this project was the equivalent of the Barcelona Pavilion.

3. The National Historic Preservation Act, Pub. L. No. 89-665, S. 3035, 89th Cong. [1966].

4. Albert Fein, "A Study of the Profession of Landscape Architecture: Technical Report," American Society of Landscape Architects Foundation, 1972.

5. According to Fritz Steiner's biographical essay on Ian McHarg in *Shaping the American Landscape* (2009), "At the urging of Russell Train, then president of the Conservation Foundation, and noted ecologist Raymond Dasman, McHarg started to pull together his theories for an ecological approach in 1966." This led to the publication of *Design with Nature* in 1969.

6.   Albert Fein, "A Study of the Profession of Landscape Architecture: Technical Report," American Society of Landscape Architects Foundation, 1972, p. 41.

7.   Ibid, p. 40.

Charles A. Birnbaum is founder, president, and chief executive officer of The Cultural Landscape Foundation (TCLF). Previously, Birnbaum spent 15 years as coordinator of the National Park Service Historic Landscape Initiative (HLI) and a decade in private practice in New York City with a focus on landscape preservation and urban design.

Chapter 12

LANDSCAPE CITY

JAMES CORNER

The past 50 years—1966 to 2016—have been an astonishing period for landscape architecture. Since the *1966 Declaration of Concern* was first penned by Ian McHarg and his colleagues, landscape architecture has experienced significant advances on so many fronts: the rise of environmentalism and ecological planning and design; the ascendancy of multidisciplinary practices, ideas, and influences; the experiments and possibilities put forward by Land Art and artists focused on landscape; the influence of serious historical reflection, design theory, and criticism, and the important cultural and intellectual positioning of the field; advances in digital and media technology, as well as the development of new materials, tools, and techniques; and, of course, the built works themselves!

Such projects are impossible to collate here, but a visit to any bookstore will show a host of exciting and beautifully vibrant landscape architectural projects from all around the world—projects more colorful, well crafted, inventive, and diverse than at any previous point in history. This is especially true for what seems like a renaissance period over the past 20 years or so, with massive investment in new large-scale urban public parks, waterfronts, reused infrastructure, public squares, trails, and other public spaces in the city. Many of these projects are multifaceted and both politically and technically complicated, with landscape architects often *leading* multidisciplinary professional teams to deliver significant transformations of cities, turning previously dormant liabilities into new and invaluable assets.

The past 50 years have brought new life and vitality to the city, and landscape continues to serve as a powerful influence, if not actual model and trope, for how the city might best evolve into the future. If the *1966 Declaration of Concern* focused on an ecological environment, I would posit that our focus today must be on the urban environment, the city—with all of its social and political dimensions, including the aesthetic and the intellectual, engaged and enmeshed within issues of nature and ecology.

Here we have our shared mandate—a new declaration that landscape architects must take on the challenges of shaping and forming the future city, quantitatively and qualitatively, ecologically and socially, pragmatically and poetically.

From a purely environmentalist perspective, the city is surely the only rational solution for continued population growth on a planet with limited and seriously diminishing natural resources. More than 75 percent of the world's population will live in cities by 2050. Higher concentrations of people in cities can help conserve precious land for water, agriculture, and resource management while reducing dependency on private automobiles with long commute times, congestion, and pollution. Cities efficiently concentrate mobility, productivity, culture, lifestyle, and value. If you love nature and countryside, choose to live in the city. Otherwise, little countryside or wilderness will remain, not to mention the insurmountable challenges of adequately watering and feeding an ever-growing population.

With more people, cities will inevitably become denser and more compact and will require new organizational development frameworks that improve mobility, efficiency, and comfort. Landscape architects are well positioned to lead such initiatives because they see the city as a kind of dynamic ecosystem—a layered composite of patches, corridors, networks, nodes, edges, and interfaces; and a city landscape of interconnected systems, pathways, and places. More importantly, unlike commensurate efforts to improve systems efficiency by planners and engineers, landscape architects can go one step further and strive to embed beauty, desire, and pleasure into the system. After all, the city is not just a quantitative problem but also a deeply qualitative one, wherein qualities of place, identity, experience, interaction, and exchange enhance a profound human sense of belonging, community, and enrichment.

People should *want* to live in cities not because they have no other option but because the city offers everything they desire: neighbors, friends, schools, restaurants, cafés, theaters, markets, shops, museums, parks, gardens, waterfronts, and, ultimately, identity. People live *somewhere*, and a sense of place counts. Life takes place in the city; it is a physical and psychological experience bound into the buildings, streets, parks, and the physical makeup of a particular place. Much of this experience derives from the public realm—some of it designed by landscape architects, often in specific ways that make places unique to a particular city, belonging to a particular place and time.

Public spaces offer exposure to locality, to nature, to greenery, and to birdsong, but also exposure to others, to the cosmopolitanism of diversity and culture, to the joys and pleasures of being among others in certain atmospheres and places. These are collective spaces for both the individual and the group; the young and the old; the escapist and the crowd; the exhibitionist and the voyeur—spaces for everybody in a colorful, open, and cosmopolitan democracy.

> The "city as garden" elevates experience and pragmatics to poetry and art.

Here we have our shared mandate—a new declaration that landscape architects must take on the challenges of shaping and forming the future city, quantitatively and qualitatively, ecologically and socially, and pragmatically and poetically.

Quantitatively, landscape architects must help improve the organizational systems, frameworks, pathways, and places that structure mobility, community, economy, lifestyle, and resources like water, air, food, and natural habitat. Landscape architects should be working on initiatives involving transportation, climate-change resiliency, infrastructure, urban agriculture, stormwater management, waste management, and urban design development structures, including organizing block types, streets, open spaces, and connective pathways. Intelligent urban structuring for the twenty-first-century city is essential for social democracy, for human health and well-being, for ecological sustainability, and for economic competitiveness.

At the same time, landscape architects must also work qualitatively to imbue the city with beauty and character and with richly varied, artfully designed public spaces—from the smallest pocket park to the urban square to the largest urban park system; these should all be places that lift the human spirit, promote open and cosmopolitan forms of social interaction, and create unique places around which people can interact, center, and belong.

Of course, the city may always be primarily motivated by economic imperatives—real estate, commerce, and profit—but as a humanistic endeavor, the city should also be a beautifully designed place for public life and pleasure. The city should be a well-organized framework that allows life to take place efficiently, productively, and comfortably, but it should also be a richly textured and idiosyncratic platform that maximizes exposure to otherness, to difference, to the unforeseen, to the wild, to the promiscuous, and to the very essence of the social. The city needs to return to being a place of both improved utility *and* irresistible desire.

Hence the virtues and lessons of the garden; in its very historical essence, the garden is a place that requires the utmost technical, pragmatic know-how and care, on the one hand, combined with the aesthetic art of design, placemaking, imagination, wonder, and invention, on the other. The city as garden elevates experience and pragmatics to poetry and art.

As we continue to talk about environmentalism, sustainability, resiliency, and so on—the mantras of the *1966 Declaration of Concern*—we must also draw from the humanistic dimensions of the landscape imagination, prioritizing beautifully designed cities and public spaces as living platforms for socialization, cultural value, and pure human experience. The twenty-first century is going to be shaped by plural, multifaceted identities, and by booming cities that celebrate density, diversity, and vibrant forms of public life and interaction. An open democracy (unlike protectionist borders) suggests the design of public spaces that are inclusive, porous, catalytic, interactional, and transactional; it suggests living spaces of encounter that are beautifully inviting, technically progressive, morally uplifting, and socially edifying—imaginative, transformative projects for a new twenty-first-century garden city.

James Corner is founding partner of James Corner Field Operations, based in New York City, a landscape architecture and urban design studio with public realm projects around the world. Corner is also professor of landscape architecture at the University of Pennsylvania School of Design. He earned a bachelor's degree at Manchester Metropolitan University (England) and a master of landscape architecture at the University of Pennsylvania.

Chapter 13

THE LANDSCAPE ARCHITECT AS URBANIST OF OUR AGE

CHARLES WALDHEIM

The anniversary of the founding of the Landscape Architecture Foundation (LAF) and the original LAF *1966 Declaration of Concern* invites us to revisit the identity and aspirations of the field itself. The founders of the so-called new art of landscape architecture specifically identified architecture as the most appropriate cultural identity for the new professional. In so doing, they proposed an innovative and progressive professional identity. This new liberal profession was founded during the second half of the nineteenth century in response to the social, environmental, and cultural challenges associated with the industrial city. In this milieu, the landscape architect was conceived as the professional responsible for the integration of civil infrastructure, environmental enhancement, and public improvement in the context of ongoing industrialization.

American boosters of the new art of landscape committed the nascent profession to an identity associated with the old art of architecture. The decision to identify architecture (as opposed to art, engineering, or gardening) as the proximate professional peer group is significant for contemporary understanding of landscape architecture. This history sheds compelling light on the subsequent development of city planning as a distinct professional identity spun out of landscape architecture in the first decades of the twentieth century, as well as on debates regarding landscape as a form of urbanism at the beginning of the twenty-first century.

This line of inquiry points toward the long-standing lineage of ecologically informed regional planning that grew out of the origins of landscape architecture in the first half of the twentieth century. That tradition evolved in the reformulation of landscape architecture as a highly technical

Many contemporary landscape designers deploy ecology as a model of urban forces and flows, as a medium for deferred authorship in design, and as a rhetorical device for public reception and audience participation.

and specialized branch of environmental science in the second half of the twentieth century. It was in part based on the sense of landscape architecture's potential as a scientific activity that many of the original declarations were framed. Over the last half century, this position has come to stand for an empirically informed planning process dependent upon a robust welfare state for implementation.

For a generation of landscape architects trained primarily as environmental advocates, this approach proved to be an unfortunate detour en route to the anticipated enlightened future of rationally informed ecological planning of urban form. In too many contexts, rational ecological planning came to be perceived, rightly or not, as anti-urban. It was equally received in many contexts as transcendentalist and ultimately rather anti-intellectual. The commitment of the identity of

the field to a subdisciplinary sphere of environmental science also came to be seen as less than pragmatic in the context of a withering welfare state and the rise of a neoliberal economy.

The recent renewal of landscape's relevance in discussions of contemporary urbanism has little to do with ecologically informed regional planning and much more to do with an understanding of contemporary design culture. Today, the challenges of urbanization are less about the strengths of empirical knowledge and scientific method and more about the political failures of a culture that has largely abandoned welfare-state expectations of rationally informed ecological planning. Landscape's recent renewed relevance to urbanism, instead of originating in the long-standing tradition of environmentally informed regional and urban planning, springs from landscape's recent rapprochement with design culture.

In many ways, the contemporary interests of the most recent generation of leading landscape designers originated within architectural discourse during the past quarter century, as if postmodernism has finally come to landscape. Not surprising, many of those leading landscape architects began their education in landscape ecology, only to have that knowledge catalyzed by architectural theory. The generation of landscape architects and urbanists trained in this way exhibits a tendency to combine several seemingly contradictory understandings of ecology. Many contemporary landscape designers deploy ecology as a model of urban forces and flows, as a medium for deferred authorship in design, and as a rhetorical device for public reception and audience participation. They also have recourse to the traditional definition of ecology as the scientific study of species in relation to their habitats, often in service to a larger cultural or design agenda. In addition to its status as a model, ecology has come to be an equally effective metaphor for a range of intellectual and disciplinary pursuits.

Ecology is relevant as an epistemological framework operating at the level of a metaphor in the social or human sciences, the humanities, history, philosophy, and the arts. This metaphorical understanding of ecology has been particularly significant as it has been absorbed into the discourse around design. While landscape architecture and urban planning have historically tended to view ecology as a kind of applied natural science, architecture and the arts have received ecology as a metaphor imported from the social sciences, humanities, and philosophy. In the most intriguing of contemporary urban projects conceived through this understanding, urban form is given not through planning, policy, or precedent but through the autonomous self-regulation of emergent ecologies. In many examples, the ultimate urban figure is attained not through design but through the agency of ecological process directed toward social, political, and cultural ends.

A survey of contemporary landscape design practices internationally offers a provisional thesis: landscape design strategies often precede planning. Frequently, ecological understandings inform urban order, and design agency propels a process through a complex hybridization of land use, environmental stewardship, public participation, and design

> Frequently, ecological understandings inform urban order, and design agency propels a process through a complex hybridization of land use, environmental stewardship, public participation, and design culture.

culture. In these projects, a previously extant planning regime is often rendered redundant through a design competition, donor's bequest, or community consensus. The landscape architect, operating as an urbanist, reconceives the urban field, reordering the economic and the ecological, the social and the cultural, in service to a newly configured urban condition. Collectively, these practices represent the landscape architect acting as the urbanist of our age.

Charles Waldheim is John E. Irving Professor of Landscape Architecture at Harvard Graduate School of Design where he directs the school's Office for Urbanization. Waldheim serves as Ruettgers Curator of Landscape at the Isabella Stewart Gardner Museum.

Chapter 14

CONNECTING HUMANS AND NATURE FOR AN IDEAL FUTURE

FENG HAN

"Mountains and rivers are texts on the desktop while texts are mountains and rivers on the desktop."

This old Chinese saying reflects Eastern thought that all human knowledge and achievements were inspired by nature and nature exists everywhere in human life.

Deeply rooted in nature, landscape architecture deals with human-nature interaction through landscape design and construction. Today, not every country provides landscape architecture as a field of study, and not every language has a word for *landscape*. However, every nation has a view of nature dating to ancient times, each with unique beliefs, wisdom, knowledge, and emotions inspired by nature. The resulting ideals and aspirations are brought into daily work and life through landscapes. Based on this connection, the discipline of landscape architecture bridges all of us to several basic understandings from past to future.

First, landscape is the representation of a view of nature. Landscape reveals human values and views of nature based on the deep-seated philosophical core of the human-nature relationship. Although we have only one objective nature, the view of nature varies by culture. Different cultural and ethnic groups show their understanding of the wisdom of nature through their own cultural filters, which drives us to innovate in the living environment. This, in turn, reflects the global diversity of the spirit and practice of landscape, making a dialogue about landscape possible among different cultural and ethnic groups.

Second, landscape is a treasured heritage of the combined work of humans and nature. Landscape embraces the wisdom of nature, creates worldly ideals, shows respect for life forms other than humans, and exercises ethics and wisdom in living in harmony with other life forms. Landscape is humanistic as well as scientific. A variety of landscapes express long and intimate relationships between people and their natural environment, reflect specific techniques of land use that guarantee and sustain biological diversity, and embody the spiritual relationship of people and nature associated in the minds of the communities with powerful beliefs and artistic and traditional customs. Landscapes are a treasured human legacy for sustainable development for future.

Third, the meaning of landscape is socially constructed. Landscape is socially constructed, so it has a layering of meanings associated with the social subjects of construction, the construction process itself, and the cultural input process, which has great social relevance to beliefs, knowledge, and the use of nature by the authors and communities. Landscape

is a social container and space to reflect deep social values, relationships, emotions, and memories.

These core values of landscapes face unexpected threats, especially because of intensified globalization and urbanization. Environmental degradation, increasing social conflicts, and the loss of harmonious human-nature relationships, local identity, and historical context become significant topics of concern for the future. Landscape architecture should play a role in advocating a historical and evolutionary holistic methodology and approach and thereby lead the way in the overall development of the built environment of the future. The relationships that humans have with nature, contested landscape, cultural landscape diversity, natural values, heritage landscape conservation, and sustainable landscape approaches and tools are key issues that need to be addressed.

Rebuild a sustainable relationship between humans and nature with a sense of urgency. Accelerating urbanization, disorderly urban expansion, and the exploitation of natural resources have exacerbated the divide between humans and nature. Pollution, smog, and food insecurity threaten the necessities people need to survive. The relationship between humans and nature has reached a critical point. Environmental philosophy should be applied to the landscape architecture framework. Only in this way can our attitudes toward nature

> Landscape is a social container and space to reflect deep social values, relationships, emotions, and memories.

be fundamentally changed and sustainable development improved. Science cannot save us from the root.

Focus on contested landscapes and create social equity in landscape. The reorganization of social spaces and population migration accompanied by globalization and urbanization are, in essence, the redistribution of natural resources and land resources. In this process, we must consider land tradition, emotions, memory, and interest. We must face contested social spaces and ensure that the new landscape becomes traditional inheritance and sustainable development, recognizing social welfare and justice instead of social and capital privilege. Landscape architects should be more socially responsible.

Reinforce understanding and sustaining landscape as cultural habitat. Each landscape is not empty. It has its own stories and history. Involving local communities, recognizing and respecting their cultural customs, and using both innovative and traditional practices can provide more effective management and governance of multifunctional landscapes, contributing to their resilience and adaptability and enhancing local identity. Landscape creation should keep cultural responsibility and historical context in mind while acknowledging individual innovations.

Enhance understanding of intrinsic natural values and respect for nature. Nature speaks its own languages. Strengthening interdisciplinary scientific work in landscape,

enhancing our knowledge of the intrinsic values of nature, and taking concrete actions in urban and rural development to conserve and interpret natural values are the keys to maintaining harmony between humans and nature. Implementing this core task of landscape for a sustainable built environment is a win-win strategy for humans and nature in sustainable coexistence.

Protect and manage landscape heritage. A bright future can be ensured only by knowing the past. Landscape heritage embodies the wisdom of people who know nature and land in a particular natural environment, their spiritual pursuit, emotional memory, and cultural identity. All are valued treasures of human society. Effective protection and management of heritage provide a great engine for future

> Our objective is not just to live intelligently and healthily, but also to live on earth spiritually and poetically together with all other forms of life.

development. We must ensure that we remember where we came from and can recognize the way home even while we move into the future.

Develop strategies and tools for sustainable landscape planning and design. There is an urgent demand in landscape architecture education and training to provide multidisciplinary knowledge and solutions. Concrete planning and design strategies, approaches, and tools that can integrate economic, social, cultural, and environmental processes inclusively for sustainable development are strongly advocated.

Our objective is not just to live intelligently and healthily but also to live on earth spiritually and poetically together with all other forms of life. That is the ambitious task for future landscape architects.

Feng Han, PhD, is professor and director of the Department of Landscape Studies, College of Architecture and Urban Planning, Tongji University, Shanghai, China. She received both her BLA and MLA from Tongii University and her PhD from Queensland University of Technology in Australia.

Chapter 15

URBAN ECOLOGY AS ACTIVISM

KATE ORFF

Nineteenth-century Olmsted parks shaped green civic spaces, forming a backdrop for an emerging democratic nation and defining the contours of the American city. Today, urban landscapes are newly pivotal in fostering an era where communities directly engage with the local environment, and where these spaces can be reimagined as productive landscapes that are not only pastoral settings but also active generators of social life. I envision an activist future landscape that gives form to citizen participation and grows in tandem with social networks. In the age of climate change, everyone is a landscape architect.

The book *Toward an Urban Ecology* (Monacelli 2016) maps our design process: exploring, defining, researching, and building these landscapes. Part monograph, part manual, part manifesto, it asks what the agency of the landscape architect

is. How do we not just make landscapes, buildings, and public spaces, but make *change?* Landscape architecture is not just a discipline, it is a *stance*—a stance of activism.

We need to imagine a wholly different set of relationships with the earth relative to nothing less than the scale of humanity and our shared path forward. The effects of chemical pollution, energy extraction, water scarcity, poverty, extremism, species extinctions, and social fragmentation intertwine in a global feedback loop.[1] We need not only the sharpened digital, design, and mapping tools of the landscape architect but also the tools of politics, science, storytelling, sharing, and collaboration to bring disparate groups together around a common purpose.

It will require Americans and other rich nations to consume less. It will require a rapid transition away from fossil fuels and toward less polluting forms of energy that occupy and affect our landscapes differently. It will require the proliferation of human rights and gender equity. It will require new land policy and legal concepts of landscape, not those formulated in the era of invisible hydrocarbon combustion. We need to abandon inherited picturesque concepts like viewshed, which interpreted in our legal system has doomed aquaculture, solar, and wind energy projects. It will demand sacrifice and new conceptions of pleasure, leisure, happiness, work, pain, time, and beauty. But landscape architects can help change the here and now and help society move toward a settlement pattern and mode of living that is both deeply joyful and deeply decarbonized.

> Landscape architecture is not just a discipline, it is a *stance*—a stance of activism.

We need to jointly conceptualize the physical and social and move past old notions of formal and informal, of maintenance and stewardship, to embrace a more complex understanding of landscape and community. Moving forward, landscape architects and urban designers can contribute to positive and purposeful civic-scale interventions, interweaving science, policy, people, and art. We need more projects like Living Breakwaters, a chain of protective breakwaters seeded with oysters, creating habitats designed to recruit finfish and shellfish and tended by high school science students. Such projects have emerged from a stance of informed creativity and an impulse to engage the world as it is, using contextual, holistic, and collaborative work processes. Living Breakwaters is in the preconstruction phase, advancing all components of the project with a tripartite purpose: risk reduction, marine habitat enhancement, and social engagement proceeding in concert. Projects of the future will be designed through different processes and will take very different forms when multiple purposes are taken as goals.

In this, as in all of our projects, we aim to overlay the regenerative capacity of living infrastructure with the methods of community organizing. Issues as massive as global climate change can feel well beyond our capacities to alter, but by bringing together large-scale strategic planning practices and community-based participatory initiatives, we can work together strategically to reverse ecological degradation and social fragmentation. The concept of civic landscape as a manageable scale of thought and action, which scales down to the unit of individual behavior and up to the frame of regional

> We need to jointly conceptualize the physical and social, and move past old notions of formal and informal, of maintenance and stewardship, to embrace a more complex understanding of landscape and community.

politics, remains full of potential. In this way, urban parks are redefined as next-century infrastructure, linking citizen participation with environmental prerogatives, cultivating a civic ethos of shared work, learning, play, and responsibility. Landscape architecture is seeding the next generation of engaged and environmentally aware civic stewards. Time to get to work.

1.   Richard Misrach and Kate Orff, *Petrochemical America* (New York: Aperture, 2012).

Kate Orff is director of the Urban Design Program at Columbia University and founder of SCAPE, a 30-person professional practice based in lower Manhattan. She holds a bachelor's degree in political and social thought from the University of Virginia and a master of landscape architecture from Harvard Graduate School of Design.

Chapter 16

LANDSCAPE BEYOND THE BIOTIC: IN ADVOCACY OF A REVISED LITANY

CHRISTOPHER MARCINKOSKI

Ian McHarg often wrote and spoke of what he termed his *litany*: the various subject matters and concerns that he believed all landscape architects should actively engage in their work. That litany forms a central tenant of the *1966 Declaration of Concern*, with McHarg and his colleagues arguing that students and practitioners "must know geology, physiography, climatology, [and] ecology to know why the world's physical features are where they are; and why plants, animals, and man flourish in some places and not in others." The declaration asserts that only once designers are able to "interpret" a landscape through these lenses are they properly prepared to plan and design the environment. The 1966 declaration, in essence, advocates that landscape architecture be understood as an applied arm of the natural sciences, ostensibly setting the intellectual orientation and interests of the discipline for the last half century.

While the logic behind this declaration is understandable, given the discipline's widely accepted origin story, not to mention the particular cultural moment in which it was written, I choose to reject it as far too narrow in scope to continue to serve as a useful disciplinary apparatus. For example, rather than what McHarg suggested would be the "emancipation" of landscape architecture, I contend that the discipline's strident subscription to biotic ecology above all else has unnecessarily shut out discussions of other essential landscape concerns, which has served to limit the discipline and prevent it from achieving a much greater and more significant cultural agency over the last half century.[1] As such, I suggest that now is an ideal time to establish a revised litany for the twenty-first-century landscape architect, one that is far more multivalent than the original declaration.

To begin, I would advocate a revision that centers less on purely environmental interests and more on developing a deeper awareness and understanding of those political, economic, and sociocultural machinations that motivate the urbanization activities at the center of the discipline's current and future work. Active engagement with the underlying drivers and impulses of urbanization is as important as endeavoring to limit, soften, or undo their effect. Such an approach serves to extend the discipline's field of engagement while simultaneously expanding its rhetorical kit and its means of evaluation.

I make this assertion as someone who is not formally trained as a landscape architect but rather as someone who came to the discipline because of what I saw as its potential capacity for dealing with urban issues. Particularly compelling is landscape architecture's latent facility to negotiate and synthesize a variety of competing and often contradictory concerns—economic, social, cultural, and yes, environmental—into actionable strategies and solutions, both physical and procedural. In this regard, landscape as a disciplinary apparatus offers a powerful set of instruments for navigating the myriad challenges related to the future urban condition if, and only if, it escapes the burden of advocating and evaluating its work solely through environmental terms, or the now seemingly de rigueur ecological terms. While such a contention may be viewed as heresy, I believe it is a necessary truth that the discipline must confront in order to preclude landscape architecture from being reduced to a technical vocation, rather than elevating it to an essential cultural project.

To substantiate this argument, I would like to sketch out three tenets upon which it is predicated.

First, we must accept that the conceptual binary of urban and nonurban that the *1966 Declaration of Concern* relied on is no longer valid. To paraphrase Henri Lefebvre, all that is affected by anthropogenic concerns is made urban and, as such, the entire landscape of the earth as we know it today, both physically and conceptually, must be understood as an urbanized condition of varying intensities. As a result, the fundamental question confronting planners and designers is no longer a matter of *where*, but rather a question of *how*. That is, how does one go about influencing, managing, or negotiating the intensity, formats, and consequences of the innumerable anthropogenic regimes that are continuously remaking the landscape of the earth? Such questions simply cannot be answered by relying solely on the natural sciences as our guide. Rather, we must

> Now is an ideal time to establish a revised litany for the twenty-first-century landscape architect....

look to the broader motivations and drivers of these regimes to uncover how landscape architecture might actively engage, infiltrate, inflect, and influence them.

Second, urbanization activities, particularly the construction of new settlement and infrastructure, have long been understood as the physical manifestation of economic growth. However, this correlation can no longer be accepted as valid. Urbanization activities are increasingly the result of speculative motivations, deployed as catalysts in pursuit of uncommon economic gain and righteous projections of political power. The result is a rapid and radical remaking of the earth's physical landscape, particularly in those polities deemed "emerging."[2] Demographic and market realities have given way to highly speculative pursuits focused almost exclusively on the physical products of urbanization, treating them as transactional instruments necessary to elevate global status. Given the artificiality of their demand, these speculative activities have a high propensity for failure—left incomplete or abandoned, made inaccessible, or suffering from pervasive vacancy—and in turn, result in both severe near-term and long-term consequences. The implication is that planners and designers can no longer train their attention solely on the preferred outcome of their work, given the increasingly low likelihood of these pursuits coming to fruition. Rather, the work must pursue active mechanisms of adjustment, modulation, and contingency, as well as the potential for abandonment or repurposing. In order for design and planning to produce these more dynamic formats of urbanization, they must be cognizant of and actively engaged with the systems and agendas motivating these urbanization activities in the first place.

Third, while it is encouraging to see an increasing number of schools and practitioners looking at expanding the discipline's topics of concern beyond the traditional scope of the landscape architect, far too much of this recent work has been centered on the wholesale replacement of existing formats of settlement and infrastructure in pursuit of novel systems conceived of almost exclusively from self-described ecological concerns. The ambition is commendable, but the potential erosion of cultural credibility that the discipline risks in promoting these kinds of projects is not. Instead of advocating naïve totalizing visions, I encourage the pursuit

Elevate other fundamental urban concerns to a commensurate status in order to move landscape architecture beyond its unnecessarily limited focus to a more multivalent, actionable, and efficacious orientation.

of strategies of inflection, appropriation, and subversion. Even though such an approach may be slower and perhaps rhetorically less dramatic, its capacity to produce real, fundamental, and established influence is far greater. To discover these moments of potential operation—locations of loose fit or slippage or surplus—it is essential to again understand the fundamental principles that have motivated the construction of these extant systems of urbanization. The natural sciences do provide one lens, but searching for potential means of intervention under these terms alone is unnecessarily reductive. However, if the field of disciplinary operation is expanded beyond its preoccupation with biotic ecology to include political, economic, and sociocultural concerns, significantly more opportunities for potential action quickly come into focus.

I do not make these assertions to diminish the environmental concerns that have typified the work of landscape architecture over the past half century. Rather, my interest is in elevating other fundamental urban concerns to a commensurate status in order to move landscape architecture beyond its unnecessarily limited focus to a more multivalent, actionable, and efficacious orientation.

If landscape concerns are to have a truly central role in the ongoing processes of urbanization being seen globally—for them to function as physical structuring devices, conceptual organizing logics, or actionable policy instruments, and not simply as tidy little neoliberal ornaments of green—they must be argued for and explained in the sociocultural, economic, and political terms of global urbanization. Only then will landscape architecture begin to approach its potential as a truly essential cultural project.

Christopher Marcinkoski is associate professor of landscape architecture and urban design at the University of Pennsylvania. He is a licensed architect and founding director of PORT, a public realm and urban design consultancy based in Philadelphia and Chicago. He holds a bachelor of architecture from the Pennsylvania State University and a master of architecture from Yale University.

1.   Ian L. McHarg, "An Ecological Method for Landscape Architecture," *Landscape Architecture Magazine*, January 1967: 105-107.

2.   Christopher Marcinkoski, *The City That Never Was* (New York: Princeton Architectural Press, 2016).

Chapter 17

FOR A LANDSCAPE-LED URBANISM

HENRI BAVA

In the face of the combined phenomena of climate change, increasing population, and the growing urbanization of our societies with their consequent economic disparities, we need profound and lasting transformation. Each of us, consciously or not, is implicated because every one of our actions has a consequence.

Landscape architects must call on their culture, their experience, and their knowledge to invent and encourage innovative solutions, to transform their actions and ways of thinking, anticipating mutations in our cities and adapting to this new context in constant evolution, using landscape, populations, and habits.

The urban projects of landscape architects—be they local or territorial—must be situated within and supported by their geography, acting within ecosystems, biocenoses, and biotopes.

The Earth

We are human beings on a planet, Earth, the same as the material with which we work. We are linked to and dependent upon this substance. It is not by chance that the words *humus* and *human* have the same origin, as humus is the source of life, permitting the soil to stock nutritive minerals, water, and carbon.

The quality of humus depends on the way it is cultivated and worked, as it is not the natural soil. It is changed by man's labor. The farmer cultivates the land before planting, maintaining, and reinforcing its living qualities—its intense, subterranean life which allows plants to germinate, develop, and flourish.

In every project, landscape architects can take inspiration from generations of farmers and support the development of humus in soils, decisive for the future of humanity. We must preserve the quality of our soils to avoid their exhaustion through the overuse of fertilizers, to keep their ability to retain water, and to maintain their porosity so that they remain open and oxygenated. Agroecology can spread at both local and global scales, with agro/urban projects, community gardens, and shared gardens. The quality of our soils, even in our cities, is as essential as the quality of the air we breathe.

> We as landscape architects must encourage everyone to become involved in maintaining the well-being of populations.

Water

Water is in constant motion, and the water we drink has already been drunk. To maintain future access to safe drinking water, we must develop alternatives to the heavy and expensive infrastructure of the past. Urban sprawl and increasing soil impermeability raise the risk of both flooding and water shortages. Among the most environmentally friendly models devised to secure the supply of water are those that produce water at a very local level, on the block or in the neighborhood, or mixed systems that retain the public water system for drinking water and have local systems for recycling graywater and rainwater for other uses.

Water in movement can bring sudden, radical changes, with associated risk for populations. We need to understand the dynamics of the water network and transform the risk into project potential, with creative responses to changing and unknown factors.

Successful water management in the metropolises must account for water quality and potability; safety and security; the environment; and resolution for the articulation of water with the city while reinforcing the ecological corridors generated by the rivers to form a landscape structure at the territorial scale.

We must recover the natural banks of rivers and their riparian forests within the city, making them visible and accessible to promote awareness and ownership by the people.

Working with the living

To work with the living is to promote diversity and to understand the ecosystems present in each project, not merely preserving the existing but encouraging a richer biodiversity while increasing the number of inhabitants.

The living is the main material—with all that implies: carbon storage in food crops, the true fuel of human beings, but also storing carbon with perennial plants, most notably trees, at the center of strategies for carbon capture and energy storage. Maintaining the biodiversity reservoirs that are the wastelands in and around cities is necessary and will mean changing our views on these rich and complex spaces.

The well-being of populations

We as landscape architects must encourage everyone to become involved in maintaining the well-being of populations—that quality of people living well together, place by place, with ownership of a project and participation in its realization.

We must avoid urban sprawl and encourage the dense and fertile city by establishing nature in the city.

Cohesion and social diversity, accessibility of services, a favorable environment for active mobility, a healthy community, a pleasant living environment—all are necessary. However, developing appropriate responses to political and social realities is not simple; it takes time. The knowledge that we are faced with this complexity, on the one hand, and the undeniable link between the quality of the environment and the health of populations, on the other hand, are worth sharing—becoming enriched through interdisciplinary and intersectoral reflection, and augmented by research.

Density, compactness, and fertility

We must avoid urban sprawl and encourage the dense and fertile city by establishing nature in the city. This is in no way a contradiction: density is not antithetical to nature in all its forms—on roofs, streets, or squares—promoting project ownership by human inhabitants, with their activity bringing life to the project.

Density is associated with the issue of mobility, transportation with little or no pollution, clean air, energy conservation, and renewable energies, and forming part of the equation for virtuous cities.

When these elements are in synergy, they create the conditions for a new paradigm in which landscape architects in particular must play their full part in the design of the cities, landscapes, and territories of tomorrow, directing design teams in the respect of sites and landscapes, to install a landscape-led urbanism.

Henri Bava is founding partner of Agence TER, with offices in Paris, Karlsruhe, and Los Angeles. He studied plant biology at the Université Paris-Sud, Orsay; scenography at Ecole Jacques Lecoq, Paris; and landscape architecture at the Ecole Nationale Supérieure de Paysage (ENSP), Versailles. Since 1998, Bava has been chairman of the Landscape Architecture Department at the Karlsruhe Institute of Technology (KIT) in Germany.

Chapter 18

TOPOLOGY AND LANDSCAPE EXPERIMENTATION

CHRISTOPHE GIROT

The Romans believed in three kinds of nature: an untouched wilderness, productive agriculture, and the garden as cultural and symbolic artifact. Today, only one kind of nature remains where humankind dominates, seeking to show both a mastery over, and an understanding of, the processes of natural creation. This scientific approach to nature is monitored, programmed, fabricated, and maintained through empirical methods of trial and error with the help of advanced cybernetics and modeling.

Redefining nature: questioning nature as fabrication

The late ethnologist Claude Lévi-Strauss gave a speech in 2005 entitled "On the Growing Difficulty of Living Together"[1]

where he reckoned that between the time when he was born in the early twentieth century and now, the world had gone from 1.5 billion inhabitants to more than 6 billion. The population had more than quadrupled in his lifetime and was still increasing logarithmically with each decade. In an ethical stand, he postulated that the disappearance of species was a fundamental breach in the integrity of creation, and hence the rights of humanity should cease as soon as they encroached on the survival of other species.

Now that we humans number more than seven billion, unbounded human growth will cause immeasurable repercussions on both the cultural and biological diversity of the planet, ultimately jeopardizing mankind itself. We have entered an age of complete natural fabrication, where landscape architects are asked to reinvent forms of nature that respond appropriately to the unwieldy challenges of climate change and rising waters. Such fabrications require a proactive understanding of ecology and diversity and will produce landscapes that challenge conventional wisdom, completely redefining the understanding of nature embedded in our collective consciousness.

Resilient design: inventing new topical landscapes

Natural fabrication will undoubtedly require sustained multidisciplinary exchanges and enough creative power to invent new forms of landscape that can match the formidable societal and environmental challenges ahead. John Hennessy, former president of Stanford University, created a cross-disciplinary joint program at the beginning of the twenty-

Now that we humans number more than seven billion, unbounded human growth will cause immeasurable repercussions on both the cultural and biological diversity of the planet, ultimately jeopardizing mankind itself.

first century that merged schools—ranging from engineering to the arts—and helped break down conventional barriers between disciplines, thus allowing creative interaction within established structures and the questioning of old habits.[2] By training generations of engineers, designers, and artists to work together in synergy, to think differently when faced with unforeseen challenges in the environment, we, too, could become more responsive creatively.

Resilient landscape design requires a significant change of attitude, through the invention of new terrain topologies that will perform better in an environment with shifting boundary conditions. New topical landscapes will emerge that better express the specific needs and priorities of a world exceeding seven billion inhabitants. Designers may observe, understand, and acquire knowledge about a particular situation, but they will also need to react and think creatively in synergy with others about solutions for the future. A topical selection of projects will suppose a fundamentally new mode of interaction between disciplines, where landscapes will be prototyped, simulated, and tested in a series of feedback loops.

This process of dialogue and interdisciplinary design iteration will undoubtedly give birth to new forms of natural systems and will have deep repercussions on the secular values that we presently attribute to nature. Landscape education should, therefore, encourage creative invention through reinforced dialogue with other disciplines, developing design tools and methods that are better adapted to our rapidly changing world. Topical landscapes produced through this exchange will challenge many preconceptions and bring renewed confidence and creative ability to our field by fostering innovation through experimentation in design.

Commonalities and divides: toward a culture of difference

Recognizing and fostering regional and cultural differences in landscapes will probably be the strongest challenge of this global age. As with the disappearance of species, countless

Recognizing and fostering regional and cultural differences in landscapes will probably be the strongest challenge of this global age.

cultural traditions related to landscape will disappear in coming decades due to the increased political, economic, and environmental pressures of globalization. French geographer Augustin Berque reminds us of the very particular relationship that humans have with their natural environment, a relationship that differs strongly and symbolically from one society to another, from one language to the next.[3] Berque defines this bond with nature as the *ecumene* and reminds us of the unfathomable differences between cultures of the Orient and the Occident, and the North and the South. The *ecumene* expresses, therefore, a society's attachment to a particular landscape reality, an ontological predisposition toward nature, where the relationship to landscape is understood as a set of strong beliefs and signifiers. Our relationship to the world is the complex product of language, work, culture, and myth, and the idealized expression of this faith in nature often merges at the cusp of strong cultural divides, where things can barely be explained, let alone be sensed. Commonalities

and environmental concerns will thus continue to face prevailing linguistic and cultural divides, nurturing strong distinctions and discrepancies between human societies. As cultures disappear, other hybrids will arise, underlining the prevalence of local lore over globalization, and each landscape will thus become an invitation to express a culture of difference in an act of superb creative defiance.

1.   The speech, entitled „*La difficulté croissante de vivre ensemble*" ("On the Growing Difficulty of Living Together"), was written and presented by Claude Lévi-Strauss in 2005 upon receiving the 17th International Prize of Catalonia.

2.   Reinhold Steinbeck, "Building Creative Competence in Globally Distributed Courses Through Design Thinking," *Revista Communica*r 19, no. 27 (2011): 27-34.

3.   Augustin Berque, *Histoire de l'Habitat Idéal: De l'Orient vers l'Occident* (Paris: Editions du Félin, 2010), 45. Berque speaks about a succession of fundamental upheavals in our relationship to the world, for instance, though Chinese Culture at the time of the Six Dynasties, when the relationship between nature and culture was dematerialized and inverted through the monastic practice of the ecumene.

Christophe Girot is professor and chair of landscape architecture in the Architecture Department of ETH Zurich. He received a dual master of architecture and landscape architecture from the University of California, Berkeley, and was chair of design at the Versailles School of Landscape Architecture. At ETH he cofounded the Landscape Visualizing and Modelling Laboratory (LVML) with Professor Adrienne Gret-Regamey in 2010.

Chapter 19

REGENERATIVE INFRASTRUCTURE SYSTEMS THROUGH URBAN ACUPUNCTURE

TIM DUGGAN

The twentieth-century transportation network created substantial challenges for many communities across the country, and we face significant aging infrastructure and economic development challenges for the twenty-first century. Together, local, state, and federal governments have set about on a frenzy of urban renewal projects through brazen demolition and bold, yet sometimes ill-advised, construction of infrastructure, buildings, and roads. The coordinated effort that built our modern highway system and provided quick access to and from a majority of urban cores had several unintended consequences that emerged over time. This approach, unfortunately, also left a legacy of isolation in its wake for many low- to moderate-income communities, with readily apparent neglect across our urban landscape. More importantly, these infrastructure systems are not sustainable and should not simply be replicated with infrastructure that is copied and pasted from the last 100 years. Infrastructure systems need to be designed with multiple positive outcomes through innovative processes, with the goal of developing more regenerative infrastructure solutions.

Having internalized these lessons of the past, landscape architects are at a crossroads where they can either maintain the status quo or elbow their way to the "decision-makers" table and help establish a new paradigm for solving this current set of problems. Our profession needs to help shift patterns in the way that city infrastructure systems are planned, designed, and operated and adapt them to how present-day residents live, work, and play. These systems must be integrated into the natural environment in a more sustainable manner.

The shift should begin at the urban core and focus on making existing blighted communities and abandoned brownfield landscapes into healthier, walkable, resource-rich, and well-connected communities. Simultaneously, the

Landscape architects need to orchestrate regenerative infrastructure initiatives that are centered around safe, healthy, walkable, digitally connected, resource-rich environments with a goal of driving systemic change within our communities and leveraging our limited infrastructure resources for the next 100 years.

profession needs to demand wholesale change in the decision-making and design process from the previously applied limited resources and funding mechanisms—from scattered sites and piecemeal system improvements into highly strategic node-based redevelopment initiatives. If we truly want to overcome blight and disinvestment in the marginalized sections of our cities, landscape architects will need to be more strategic and creative with our approach to designing infrastructure systems across the country. This planning approach to creating regenerative infrastructure systems that can leverage multiple community benefit outcomes has been referred to as *urban acupuncture* by Jaime Lerner, Bob Berkebile, and others.

Streets once designed for horse carts must now safely carry cars, buses, freight, digital utilities, stormwater, wastewater, cyclists, and pedestrians. In areas where growth has yet to arrive, the existing transportation infrastructure fails to provide nodes and corridors that are sustainable into the future. We can no longer let the automobile make every decision in the built environment.

Landscape architects need to orchestrate regenerative infrastructure initiatives that are centered around safe, healthy, walkable, digitally connected, resource-rich environments with a goal of driving systemic change within our communities and leveraging our limited infrastructure resources for the next 100 years. This broad understanding of systems-based thinking can empower the profession to coordinate large-scale infrastructure opportunities like transit-oriented development, affordable housing, and mixed-use development, as well as leverage those limited resources into very strategic urban revitalization outcomes.

Landscape architects must take a unique approach in how we seek ways to collaborate, engage, design, and leverage these existing opportunities into regenerative infrastructure solutions. These projects have the opportunity to redefine infrastructure in America for the next 100 years. By harnessing all the opportunities of these projects, a new vision

We know from experience that big visions are actually easier to accomplish than their less popular cousin—the "not-so-interesting vision." However, big visions have to be grounded in a deep understanding of physical and market constraints with real data.

can be realized with the collective experience in planning, designing, and implementing sustainable solutions locally and across the country. The diversity of project experiences in these infrastructure systems makes our profession uniquely qualified to be the leaders in those efforts.

This approach can only be accomplished through a highly collaborative process rooted in technical understanding, community engagement, and stakeholder involvement. The broad understanding of systems-based design and collaboration is the heart of landscape architecture, a profession that can efficiently help redefine the urban landscape and assist in the development of effective, transformative models for regenerative infrastructure and public landscapes across the country.

We know from experience that big visions are actually easier to accomplish than their less popular cousin—the "not-so-interesting vision." However, big visions have to be grounded in a deep understanding of physical and market

constraints with real data. This holistic approach will form a complete solution and encourage the fusion of regenerative infrastructure systems for the next generation that is uniquely its own.

I understand that these project types present unique opportunities and constraints to redefine this paradigm within our profession and I am a willing partner in that collaboration.

Tim Duggan is founding partner of Phronesis with offices in Kansas City and New Orleans. Phronesis is a landscape architecture and urban design studio focused almost entirely on creating regenerative infrastructure and community systems within the public realm.

PART IV

THE FUTURE OF THE DISCIPLINE

AS DESIGNERS VERSED IN BOTH ENVIRONMENTAL AND CULTURAL SYSTEMS, **LANDSCAPE ARCHITECTS ARE UNIQUELY POSITIONED** TO BRING RELATED PROFESSIONS TOGETHER INTO NEW ALLIANCES TO ADDRESS COMPLEX SOCIAL AND ECOLOGICAL PROBLEMS. LANDSCAPE ARCHITECTS BRING DIFFERENT AND OFTEN COMPETING INTERESTS TOGETHER SO AS TO GIVE ARTISTIC PHYSICAL FORM AND INTEGRATED FUNCTION TO THE **IDEALS OF EQUITY, SUSTAINABILITY, RESILIENCY, AND DEMOCRACY**.

FROM THE NEW LANDSCAPE DECLARATION

Chapter 20

INTO AN ERA OF LANDSCAPE HUMANISM

GINA FORD

The landscape architect is uniquely rooted in the natural sciences. He is essential in maintaining the vital connection between man and nature.

—*1966 Declaration of Concern*

Fifty years ago, the voice of our profession was eerily prescient, undeniably smart, and powerfully inspired. It was also, let us admit it, almost entirely white and male.

I note this with no disrespect to the six incredible leaders of our profession who penned the *1966 Declaration of Concern*. Their call to reconcile the needs of humankind with sound knowledge and respect for the natural processes of our environment is as relevant today as it was then or even more so. Equally, their edict for landscape architects to command the technical skillsets associated with natural resources and

processes (geology, physiography, climatology, and ecology) remains of vital importance.

Yet, as we look forward and consider the significance of climate change, demographic shifts, and income inequality, the declaration's "man" as nature's antagonist feels strangely abstract and incomplete. To maintain relevance over the next 50 years, the profession needs to demonstrate the highest level of natural systems expertise but must devote equal attention to the human dimension of the equation. Accordingly, this response outlines some reasons why we as a profession need to diversify our ranks, as practitioners need to sharpen our technical skills vis-à-vis the needs of the people we serve, and as collaborators need to hone our communication skills and leadership prowess.

Our profession: diversify our ranks

The *1966 Declaration of Concern* calls for recruiting and retaining more trained landscape professionals to help fight the environmental crisis. Yet broadening our social impact over the next 50 years will also require diversifying ranks and retaining population segments that have historically seen slim representation or significant attrition. Our national population demographic is changing dramatically, but our profession's membership does not correlate with this trend. Between 2000 and 2010, the Hispanic and Latino population increased by 43 percent, and the African-American population by 12.3 percent, but together these groups represent less than 10 percent of the membership of the American Society of Landscape

Fifty years ago, the voice of our profession was eerily prescient, undeniably smart, and powerfully inspired. It was also, let us admit it, almost entirely white and male.

Architects (ASLA). Moreover, while many graduating classes from the country's best schools of landscape architecture are predominantly female (and have been for a number of years), the percentage of women in leadership positions in private practice is still far from representative.

These imbalances are certainly not unique to landscape architecture, and more and more are seen across industries as both a socio-ethical challenge and a barrier to achieving more innovation. In general terms, the business case for diversity includes the cost savings associated with the retention of talent, the value of mirroring the profile of the marketplace, and the quality of ideas generated by less homogeneous teams. In design terms, our ability to stay relevant as a profession relies on our desire to reflect the broader population and embrace more diverse perspectives and approaches.

In practice: design with humanity

The *1966 Declaration of Concern* points to a series of impending environmental disasters and laments the potential for "life in such polluted environments" to become "the national human experience." I have personally worked in a number of communities affected by recent natural disasters, including floods in the Midwest, hurricanes on the Atlantic shore, and oil spills along the Gulf Coast. The hard fact is that many of our most socially vulnerable populations experience these heavily affected, polluted environments daily. More often than not, the communities most affected by natural disasters are those that are already unwittingly living with the highest levels of risk, the fewest resources for recovery, and the most underrepresented voices in broader planning dialogues.

From New York City to New Orleans, designers and planners who focus on resilience have come to realize that addressing significant environmental change demands tackling issues of social equity. We need landscape architects who understand social networks and cultural fabrics to both inspire participation in this critical dialogue and build governance toward lasting, meaningful, and sustainable change.

Collaboration: cultivate an ecosystem

The *1966 Declaration of Concern* urges "a new collaborative effort to improve the American environment." Truly sustainable development requires careful orchestration of complex layers of technical expertise as well as the inclusion of many distinct voices and constituencies. Now, more than ever, landscape architects are taking the lead on highly complex and multijurisdictional problems. Our unique ability to understand interconnected natural and human systems positions landscape architects as leaders among interdisciplinary teams and positions design as the key link among the planning, social science, and engineering disciplines. This undertaking requires exceptional communication skills from its leaders, who must translate these planning, science, and engineering concepts into legible and inspiring materials for shared evaluation and understanding.

From the reclamation of the Los Angeles River to the post-Sandy regional response led by Rebuild by Design,

> From New York City to New Orleans, designers and planners who focus on resilience have come to realize that addressing significant environmental change demands tackling issues of social equity.

landscape architects are truly in the lead on some of the complex problems of our time. As we tackle these challenges, it is our softest skills—communication, leadership, empathic listening—that will enable successful engagement and positive design outcomes.

The future of our profession is bright. The world needs us now perhaps more than ever before. We must leverage this moment and bring the best of our nation's talent to bear. If we embrace humanism as a core value in our profession, we truly can overcome the seemingly herculean challenges we face. It will certainly require understanding the specifics of the processes of our physical environment. We will also need to wholeheartedly embrace the rich diversity of who we are and strive more ambitiously to understand and meet the needs of the people we serve.

Gina Ford is principal, landscape architect, and chair of Sasaki's Urban Studio. She holds degrees in architecture from Wellesley College and landscape architecture from Harvard Graduate School of Design.

Chapter 21

ON THE FUTURE OF LANDSCAPE ARCHITECTURE

CARL STEINITZ

Size and scale matter. Landscape architects have always worked across a broad range of project sizes and scales from a small garden or a house on a difficult site, to a typical midrange of new residential areas or large parks, to regional urbanization studies or conservation strategies. The reality is, however, that most landscape architects work on smaller projects.

This has not always been the case. In the nineteenth century, gardeners made gardens and architects designed buildings, developments, and even parks. Engineers designed infrastructure for expanding towns, parks, and water supply projects. There was opportunity for those with appropriate skills to work across size and scale. The landscape-oriented professions began to coalesce around the term *landscape architecture*. Frederick Law Olmsted, Peter Joseph Lenné, John Claudius Loudon, and Warren Manning all worked across size and scale. They all designed villas and gardens. Olmsted designed a rose garden and oversaw a forest management plan for the largest private estate in America. Lenné made designs for Potsdam and Berlin. Loudon made a plan for the London region. Manning designed both gardens and the entire United States as published in *Landscape Architecture Magazine* in 1923.

I see the profession mainly, but not only, from an academic perspective. The Department of Landscape Architecture in the Harvard Graduate School of Design (GSD) was founded in 1900, aided by the Olmsted office. From its earliest days, Harvard's landscape architecture faculty taught across a wide range of scales. However, it rarely emphasized each scale equally. There has always been competition within the

Size and scale matter. Landscape architects have always worked across a broad range of project sizes and scales from a small garden or a house on a difficult site, to a typical midrange of new residential areas or large parks, to regional urbanization studies or conservation strategies.

GSD and the larger university for academic turf. This has implications for size and scale.

By the 1920s, commissions came from the wealthy needing help in designing their villas and gardens, and professional education began to shift toward smaller projects. This loss of interest in the larger landscape and the increasing (and, in my opinion, artificial and harmful) separation of design from planning caused a political division within Harvard's Department of Landscape Architecture, resulting in the creation of America's first Department of City and Regional Planning. Both departments continued to emphasize physical design as taught in studios.

In 1966, when I first joined the Harvard faculty, Charles Harris and Hideo Sasaki led an attempt to bridge the gap between landscape architecture and planning. My first teaching assignment was in a collaborative studio with Charles Harris and Reginald Isaacs, with Charles Eliot II, Phil Lewis, and Ian McHarg as visitors. This kind of collaboration continued well into the 1970s.

In the late 1970s, the planning department moved away from physical design and focused on economic analysis and social planning. It subsequently moved to the Kennedy School of Government. (Planning with an emphasis on design was reestablished at the GSD in the 1990s.) Landscape architecture focused mainly on small and midsize project design. An exception was much of my own work, extending across scales but emphasizing change in the larger landscape.

In the 1990s, the GSD made a revolutionary decision. Any advanced student could take any of approximately 15 optional studios offered each semester across the whole GSD: architecture, landscape architecture, planning, and urban design. This encouraged some studios (mine included) to address a wider range of scales. At the same time, architects were becoming increasingly interested in parks and gardens, partly because of the success of the Barcelona world's fair. Our architecture department began to offer studio projects similar to ones that had traditionally had been offered in landscape architecture.

Today, most landscape architects spend most of their professional energy working on relatively small projects. A few extend their work to include large projects. Very few work professionally at a regional scale.

Given this background, I will now try to answer the Landscape Architecture Foundation's (LAF) question about the future of landscape architecture. I think that whatever we declare, in a generation or less, things will be different. I do not believe that just because you can design at one size and scale you can design at all sizes and scales. Methods and skills over a range of scales in landscape architecture are related but different.

I think that there are four reasonable prospective possibilities.

The first possibility is simply a continuation of what we are doing today (or trying to in a changing world).

The second acknowledges competition for what has become hot property: the landscape. Architects may increasingly design landscapes as part of their architecture; urban designers may increasingly design landscapes at mid-scale; geographers, engineers, and ecologists (and politicians, bankers, and lawyers) may increasingly make design decisions for the larger landscape. These trends will impinge on education and practice in landscape architecture. External competition will likely produce a narrower landscape architecture profession.

The third possibility is perhaps the most probable and the most wrongheaded. The artificial division between planning and design will likely continue and even be reinforced, and landscape architecture will itself choose to focus on designing smaller projects. This may result from competition from professions that consider large issues such as food and water supply, biodiversity, cultural heritage, population, and climate change as shapers of landscape. Mainly, however, it will result from two factors: society's unfortunate but common caricature of landscape architecture as closely related to gardening and the landscape architecture profession's own professional and academic choices.

The fourth possibility relies on embracing the wisdom of the founders of the profession. First, we must know something that other professions do not. We must be rooted in the landscape itself and at all sizes and scales: in climate, geology, hydrology, ecology, vegetation, history, perception, et cetera. Second, we must understand that almost everything

> The artificial division between planning and design will likely continue and even be reinforced, and landscape architecture will itself choose to focus on designing smaller projects.

we do requires collaboration with other professions and with decision makers. This was the vision of my department when I joined the faculty. It is a perspective I hold to this day.

We cannot argue that we are the sole bearers of wisdom as it pertains to the landscape, and we cannot argue that we have some self-defined right to be its stewards, to be in charge, or even to coordinate design for change. Sometimes we will lead, and sometimes we will not. But if we are to work on society's most important needs, both specialist knowledge and skills of collaboration in designing are essential. We must not allow the profession to become limited in the scale and scope of its competences.

Carl Steinitz, PhD, is Alexander and Victoria Wiley Professor of Landscape Architecture and Planning Emeritus at Harvard Graduate School of Design, and honorary professor at the Centre for Advanced Spatial Analysis, University College London.

Chapter 22

CITIES FOR PEOPLE AND CITIES FOR THE PLANET

BLAINE MERKER

In 1966, Americans were abandoning central cities and leaving a trail of pollution and landscape destruction as we built into the countryside. But 50 years on, a new era has begun and we have turned back to many of those same cities. How we design them now holds the key to two of the most important questions humans face: how to live sustainably on our planet and how to be happier doing so.

Cities have unlocked a secret in the last decade: the actions that are most sustainable and even healthy are the most enjoyable as well. They let us use our bodies, they give us greater freedom, they connect us to other people, and they fill our time with things we would rather be doing anyway.

Cities give us access without the burden of ownership—access to mobility, to great amenities, to experiences. They are the ultimate innovation in resource efficiency.

Cities are experiments, and lately the distance between citizen and plan has shortened as guerrilla bureaucrats use tactical interventions that make change quicker and more nimbly. At a time when national politics are polarized and international agreements take decades, the creativity and economic might of cities have given them the resolve and the clout—more than any other part of society—to take bold environmental action. Landscape architects play leading roles in this urban story, providing innovation and inspiration by fusing human and natural processes.

But cities offer something else as well: a celebration of the human condition. Good urbanism is humanism made physical. It is a celebration of the prosaic things that make us happy. Cities let us move freely using our own bodies, they engage our senses, and they bring us together physically and in spirit. It is no wonder that cities are where talent and opportunity collect, but some are in such demand they have become exclusive enclaves. In fact, human-scale urbanism— the thing that makes cities livable—has become so coveted and is in such short supply that Silicon Valley runs its own commuter lines back to San Francisco because so many of its workers prefer to live in the city. We know from economists that income-diverse neighborhoods with short commute times create greater intergenerational economic mobility. But

> Good urbanism is humanism made physical. It is a celebration of the prosaic things that make us happy. Cities let us move freely using our own bodies, they engage our senses, and they bring us together physically and in spirit.

that opportunity is being eroded because we now have a supply problem: not enough good and green urbanism to go around. Those who cannot afford it are pushed to a kind of resource-intensive landscape that not only threatens our planet but also does not provide most of what people are increasingly looking for.

The energy-intensive, dispersed, and functionally separated "modernist" city is the great ecological experiment of the last half century. It requires people to travel long distances for daily routines. Pleasant isolated environments have been created in this experiment, to be sure, but access to collective goods and social connection here is a fragile privilege, propped up by highly subsidized individual mobility.

Along with the rise of the modernist urban form, America has become increasingly divided into two visions of how to live on the land and with one another: one dispersed and resource intensive, the other increasingly connected and efficient. This divide is mirrored in our culture and in our politics to such an extent that it sometimes seems like Americans just miles apart see the world very differently.

Landscape architecture's historical role in the modernist experiment has been to hide, bandage, decorate, and soften the profound impact of dispersal, resource-intensive transportation, and fundamental lack of human scale. It is an uncomfortable truth that our profession has often been beholden to antihumanist economics. While we protect health and safety, the risks that landscape architects mitigate are often the direct result of an inhumane urbanism operating at a scale just out of our reach.

Our experiment has exacted a sacrifice: the disproportionate use of resources, energy, and time. Compare the footprints of Atlanta and Barcelona, two cities of roughly five million people. Atlanta uses 25 times more land and produces more than 10 times the carbon dioxide (CO_2) than Barcelona.

The transportation sector is not the only issue. Additional design problems arise when we must then manage stormwater from acres of pavement or try to make mechanized landscapes safe for kids to walk to school. These are, frankly, problems we chose and which now consume much more of our design attention than they should.

The final accounting of our landscape choices is being done in our climate system, though the urgency of this, too, depends on which Americans you talk to. Where climate change is seen as a threat corresponds to the fault lines in our politics, culture, and our built environment. The kind of place we live, its political culture, and its built form, determine what we believe about our global climate system. The people who have the most at stake in dispersed, energy-intensive landscapes are the least likely to believe a problem exists.

We do not know exactly how Americans will be affected by climate change. We do know that the dispersal, separation, and mechanization of our landscape cause an experiment in CO_2 not seen in over a million years. Is it affecting us yet? The year 2016 was the warmest year on record globally. Warming is not happening everywhere equally. One place that is cooler is the southern tip of Greenland where its glaciers are turning to liquid and flowing into the Atlantic.

Our charge is not just the safety of individual people but the safety of humankind. Our envelope of best practices simply is not wide enough to address the real scale of the challenge— we need a new mentality that radically refocuses our vision.

Our changing climate is a warning that the experiment we have been running not only threatens the stability of our global ecosystem but is actually leading us further from happiness. The United States has the third highest commute times on the planet, behind only Bulgaria and Hungary. And while the option to commute is tied to economic opportunity, commutes beyond 20 minutes correlate with increasing personal misery.

And it is easy to see why. Mobility takes time and our time is limited. When we separate destinations too much, we displace

Landscape architecture's historical role in the modernist experiment has been to hide, bandage, decorate, and soften the profound impact of dispersal, resource-intensive transportation, and fundamental lack of human scale.

nourishing mobility—the kind that makes us feel excited and free—with a "junk mobility" that makes us feel bored, trapped, and stressed out. We have less time for things that make us happy: exercise, relationships, and building community.

Junk mobility is also making us broke, requiring more investment than it pays back. Expensive highways subsidize far-flung trips for low-value economic activity. They sap central cities of the residents who could support a more efficient, compact transportation system. This is not a call to divest from infrastructure but rather to consider carefully the long-range future our infrastructures lock us into. The more

There is much work to do across all sectors of society, and landscape is but one part. But without *place* as an organizer and without cities leading the way, other solutions— such as alternative power sources and better resource management—will fall short.

energy our landscapes require, the less freedom we have to solve problems other than managing energy.

Four design principles for the urban landscape will move us toward a future that makes us happier, safer, richer, and will at least give us a shot at minimizing catastrophic changes to our climate system.

First, celebrate the human dimension. A city built for people provides texture, delight, and usability for all ages and abilities. It is messy and its dimensions are smaller than most of what we build now. Human bodies, in all their idiosyncrasies and limitations, should be the client for every project.

Second, connect people better and concentrate human activity into places that create vitality and convenience. Connecting with one or two modes of transport is not enough—multiple systems using minimal energy are required for real choice and freedom.

Third, give future generations alternatives. If we do not build the cities our children and grandchildren need, they will have to use their own resources to rebuild what they inherit. Do not lock them into costly redesigns. Before sinking millions into parking garages that cannot be retrofitted or into new schools miles from neighborhoods, give the future generation—whose values will likely be different from our own—room to maneuver.

Finally, build common ground by creating places that draw diverse populations together. We found that more than half of the users of New York's street plaza program said they had gotten to know more people because of these spaces. People making under $50,000 were even more likely to say that. This is a model for how we can use place to create opportunity and connection and to bridge divides in our culture.

The good news is that the things that make us happiest— like deep relationships, human-scale environments, and a sense of kinship with those with whom we share place—are practically carbon neutral. Deep sustainability and happiness reinforce each other at every planning and design scale.

There is much work to do across all sectors of society, and landscape is but one part. But without *place* as an organizer and without cities leading the way, other solutions—such as alternative power sources and better resource management— will fall short. Landscape architects' expertise, and more importantly their values, are essential. There is much work to do, and the moment to do it is now.

Blaine Merker heads the San Francisco office of Gehl, an international design consultancy based in Copenhagen. Merker earned his master of landscape architecture from the University of California, Berkeley, where he also teaches, and holds a bachelor's degree in history from Reed College.

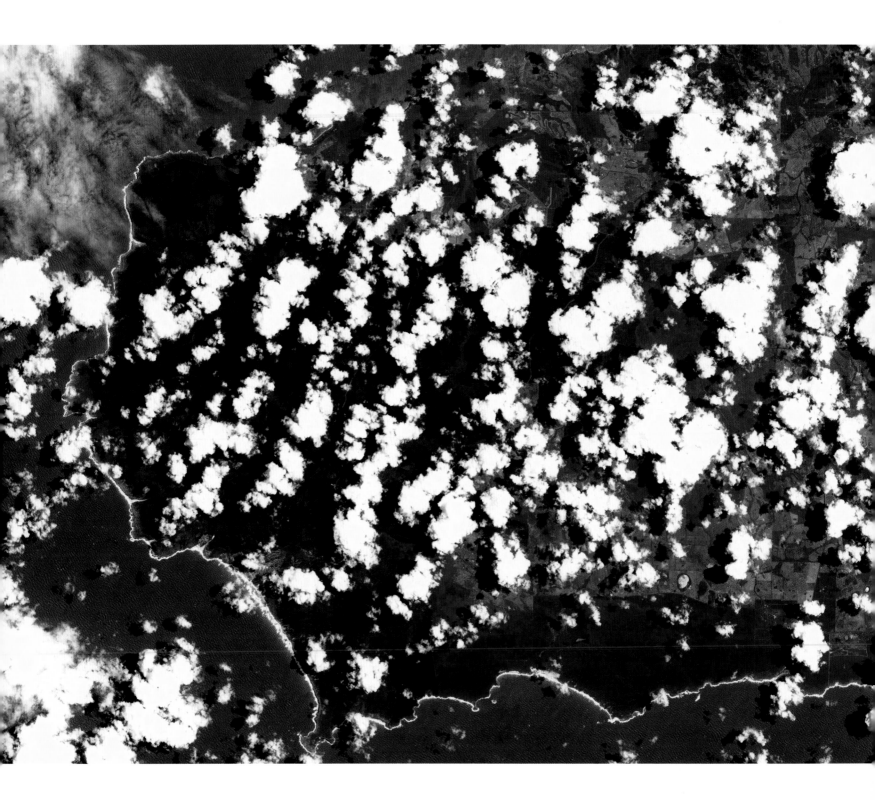

Chapter 23

THE DECLARATION OF INTERDEPENDENCE: LIFE, LIBERTY, AND THE PURSUIT OF SUSTAINABLE HAPPINESS

RANDOLPH T. HESTER

Today and every day, we reaffirm our interdependence. We offer gratitude to those prophets who declared interdependence before us: from the ancients Isaiah and Buddha to Harriet Tubman, Rachel Carson, Stewart Udall, Grady Clay, and Karl Linn, among others. In their honor, we acknowledge our responsibility to make places for life, liberty, and the pursuit of sustainable happiness. We believe that this can only be attained in the foreseeable future through an ecological democracy, a participatory government driven by systemic ecological thinking. This ecological democracy has been in the making for 50 years, 250 years, and 250 million years, evolving in mutual dependence from sea slime to the landscape of humankind. We can see our choice as living happily within our limits or perishing as ecological illiterates. We choose to bend this evolution toward sustainable happiness by committing our every resource to the following actions.

Community. That land is a community is the basic principle of ecology. So it has been for eons. Today all humankind is tied together in a single garment of destiny; whatever affects one community directly affects all indirectly. We honor the prophecies of Aldo Leopold and Martin Luther King, Jr. We commit to enhance both ecological and cultural community as one indivisible unit through each of our design actions.

Transactive language. An ecological democracy demands a new civic language to elevate discourse, to allow citizens and designers to work together, and to enable the citizenry to make decisions informed by ecological science and native wisdom. This language will be as lofty as life-affirming symbiosis and civic duty, as provocative as proactive coastal retreat and living

> An ecological democracy demands a new civic language to elevate discourse, to allow citizens and designers to work together, and to enable the citizenry to make decisions informed by ecological science and native wisdom.

smaller, as explicit as the quarter-mile radius as the limit for pedestrian trips and the scare distance for birds at rest. We pledge to create this language.

Nature. The natural landscape speaks in ways older than words, telling us all we need to know to survive, filling us with every sensual pleasure we need to thrive. Nature is the door to ecological understanding, the former experienced, the latter abstracted into principles of proper action. Even those unconcerned about the crises we face love nature. Our skills are essential to create a broad-based constituency for dramatic ecological action. We will not stand silent to witness the last child in the woods. We will make nature accessible to every child in everyday life so that each one has an equal opportunity to be eaten by a mountain lion; to watch a swallowtail emerge from its chrysalis; to skip rocks on flat water; and to build imaginary cities from mud. Such experiences give birth to a constituency for ecological democracy.

Myths. Long-held American myths must be debunked. These include, but are not limited to: the legal misconception that corporations are people with the rights of individual citizens but with few of their responsibilities; that corporate money is entitled to buy and sell our democracy and destroy our environment; that land is foremost real estate, not a community; that beauty is an elite domain, defined by professionals—a status object to be sought by others; that the landscape is flat or safely made so; and that technology can solve everything. We commit to slaying these misconceptions and never contributing to them.

Capital. Virtual capital invested by the push of a single key on a laptop invariably maximizes short-term profits by disrupting ecosystems and cultures in distant places—out of sight, out of mind. Global economies consistently ignore the value of natural systems. We commit that we will not be party to this; rather, we will help develop place-based capital and work for place-devoted clients, responsible to the communities in which their capital is invested. This requires more long-term regional practice and fewer short-term consultancies spread thin around the world.

Just beauty. The landscape can never be more beautiful than it is just. We vow to make the world both beautiful and just simultaneously. We claim environmental justice as our cause, knowing full well the difficulties therein. We commit to equitable access and distribution of landscape resources— their costs, benefits, and joys. We will create places that welcome all segments of society, not just the privileged few. We will provide open space equally in every neighborhood. We will locate necessary but undesirable land uses fairly, regardless of race or class. We will map injustices in our communities, set strategies, and find partners to overcome the injustices.

Skills. The unique skills of landscape architects built the foundation for ecological democracy and are the basis on which we offer these transformative actions. Every imminent danger is an opportunity for our holistic creativity to advance a plan based on a logical ecological unit rather than a political boundary, to design with resources others view as worthless, to make places that touch people's hearts. We will maximize seemingly mutually exclusive oppositions rather than compromise them. We will violate boundaries of discipline and scale. We will employ our expertise to act courageously on convictions. This requires our traditional skills, emerging transactive skills, and the capacity to achieve radical transformative landscapes never before imagined. We will enable the next generation with each of these skills.

Politics. Every design action is a political act. Whose politics will we style? We call for disassembling the politics in whose morass we are entangled and immobilized. We envision democracy grounded in the idiosyncrasies of local landscapes, expressing the most noble human values and distinctive ecological prognostications for each locality. Strong and visionary federal, regional, and neighborhood governance

> We will employ our expertise to act courageously on convictions. This requires our traditional skills, emerging transactive skills, and the capacity to achieve radical transformative landscapes never before imagined.

needs to subsume ineffective levels of government. We commit leadership in this overhaul. We will not tolerate the status quo. We will disobey unjust and ecologically unsound laws. We will envision, negotiate, facilitate, and provoke, each in necessary season.

These are landscape architects' actions for interdependence. Through these, we enter a gate from which we lead our society along the pathway to life, liberty, and the pursuit of sustained happiness.

Randolph T. Hester is a founder of the modern Participatory Design Movement in Landscape Architecture, professor emeritus at the University of California, Berkeley, and director of the Center for Ecological Democracy.

Chapter 24

MANIFESTO ABOUT THE PROFESSION'S FUTURE

MARTHA FAJARDO

The major dynamics of the twenty-first century—global urbanization, natural disasters, and climate change—all involve landscape architecture. They are interrelated, and landscape architects can and must address them with scientific knowledge, holistic perspectives, and creative imagination. We stand at a critical moment in Earth's history at which we must choose our future. We are living in a time of intense change with an amazing revival taking place as society, governments, and stakeholders begin to appreciate the true value of the landscape.

The adoption of the European Landscape Convention, the proposal for an IFLA/UNESCO[1] International Landscape Convention, the UNESCO Recommendation on Historic Urban Landscape (HUL), the Florence Declaration on Landscape 2012, the Latin American Landscape Initiative (LALI), the Canadian Landscape Charter, the Asia Pacific Region Landscape Charter, and the African Landscape Charter have established landscape as a vital component of collective well-being; they have highlighted the need for its management at all scales throughout the regions, including urban and suburban, cities and towns, and areas with especially degraded everyday life, as well as places of highly valued heritage and natural significance.

Landscape is a shared vision in which, and to which, a vast array of disciplines converges and makes contributions. Such initiatives are gaining interest and support around the globe. We may have different approaches to landscape, and each culture and community may understand it in a different way, but landscape is a crucial component of all our daily lives.

The fact that landscape represents the direct experience of

We are uniquely positioned to help protect existing ecosystems, to improve constructed human ones, and to regenerate those that were lost or damaged.

people in their day-to-day lives explains the growing interest in the landscape. People increasingly envision their local landscape as an engine for development and a way to boost self-esteem, identity, and quality of life.

Today we speak of landscape as a system of ecological services, as an expression of social relations, as an everyday experience, as a balance between designed open space and wild areas, as layers of meaning and values expressing life in a particular site. And this is all central to the identity of each place.

Nowadays, Latin America is brimming with new ideas and solutions. The landscape symbolizes a coming together of the natural world, human society, and people's needs. Cultivating these initiatives demands a new type of professional. We champion the landscape and the landscape profession through advocacy and support, to inspire friendly and sustainable places where people want to visit, live, work, and feel they belong. For this reason, we need to work on the significance and power of the landscape and the need to accredit suitable university courses, to promote professional development, to ensure that landscape architects deliver the highest standards of practice, and to promote initiatives to foster international, regional, and local recognition of landscapes.

Landscape architects in Latin America are especially focused on working holistically on the social component. We must bear in mind and be aware of the identity of the people who shaped the Latin American soil. These people are diverse; they originated in diverse landscapes, which should house them and provide them with proper working conditions. People should be able to use their environment.

Landscape architects are closely linked to the design of environments in which human life elapses. Therefore, we are uniquely positioned to help protect existing ecosystems, to improve constructed human ones, and to regenerate those that were lost or damaged. We can do this with thorough, thoughtful intervention. We must cultivate professional recognition in order to continue with our task.

Landscape architects do not receive nearly as much recognition as urban planners usually do. It is now the time for landscape architecture to stand with a meaningful proposal that is responsive to our present challenges.

In the past decade, landscape has been the model and medium for the contemporary city. A wide range of alternative urban practices has emerged across the world. Many of these practices explore the ecological, cultural, and territorial implications for urban changes.

Our profession has increasingly evolved away from aesthetics, garden clubs, and parks into larger urban and rural scales. Sustainable infrastructure, community well-being, landscape resilience, and social ecology are core themes for the practice of contemporary landscape architecture. Through novel projects and new narratives, in Latin America we have ratified the role of the landscape architect in a more social and human-centered approach.

Through urbanism and landscape design, cities can be transformed to reach the level that present society demands. By adopting an inclusive and innovative approach to urban renewal, cities can achieve what they have long struggled for: creating a strong culture toward change.

Culturally sensitive landscape development acknowledges diversity and promotes the ability of individuals to participate freely in cultural life, in gaining access to cultural assets as well as building a culture for "living together," helping to prevent tensions and confrontation. Thus, landscape architecture contributes to peace, conflict prevention, and reconciliation.

Landscape design has become more than greenery that heals decades of violence and fear. It is a strategically designed tool that opens civil society both physically and metaphorically.

The cities and landscapes of the future must be increasingly resilient and adaptable to changing environmental influences. Latin American natural disasters have shown the vulnerability of cities and villages. They also have proved the ability of

> Sustainable infrastructure, community well-being, landscape resilience, and social ecology are core themes for the practice of contemporary landscape architecture.

humans to exacerbate the magnitude and intensity of man-made natural hazards. Postdisaster reconstruction provides an immense opportunity for landscape architects to enhance resilience, adaptability, and regeneration of their environment.

Thus, through our praxis, we have the opportunity to make a difference by creating resilient landscapes, affordable landscapes, and landscapes of happiness, where people are the most important resource, allowing us to build for happiness and well-being.

In synthesis, our work as landscape architects is about: acting now and tomorrow; connecting nature and culture; linking practice and landscape policy; influencing landscape change toward sustainability and resilience; expressing creativity, heritage, and a sense of belonging, knowledge, and diversity; addressing a transversal and crosscutting concern and, as such, affecting all the dimensions of development;

and teaching people awareness of the meaning and value of landscape architecture.

1.   IFLA is the International Federation of Landscape Architects and UNESCO is the United Nations Education, Scientific, and Cultural Organization.

Martha Fajardo is a Colombian architect and landscape architect with a master's degree in landscape design and an honorary doctor of letters (DLitt) from the University of Sheffield. She is cofounder and chief executive officer of Grupo Verde Ltda (GVL), a professional practice in landscape architecture, landscape urbanism, and urban design. She is also cofounder and chair of the Latin American Landscape Initiative (LALI).

Chapter 25

LANDSCAPE ARCHITECTS AS ADVOCATES FOR CULTURE-BASED SUSTAINABLE DEVELOPMENT

PATRICIA O'DONNELL

For the *New Landscape Declaration*, I seek to provide a global context, motivation, and avenues for our collective actions toward a hopeful, equitable future. For three decades, I have participated on the front lines, with national and global colleagues, addressing cultural landscapes and urban landscape heritage. These experiences have revealed a key truth: you must show up and speak effectively to contribute.

Six colleagues authored the Landscape Architecture Foundation's *1966 Declaration of Concern* focusing on environmental concerns and the future of our profession. Times change, and today's complex concerns pose a multitude of challenges and the skill sets of landscape architects can contribute to solutions.

Sustainability—a larger construct integrating society, economy, and environment—emerged in *Our Common Future* (WCED, 1987). Landscape architects have both embraced and disdained sustainability, with work that integrates these aspects or work that, obsessed with form and aesthetics, emphasizes alternate values and marginalizes our impact.

The record reveals that we missed opportunities in the 1990s to demonstrate the relevance of our profession to the earth and humanity.

I recall a global rumble rising at the millennium, and in response, world leaders agreed to the United Nations Millennium Development Goals (MDGs). Those eight goals addressed poverty, hunger, gender equality, health, environment, and development with an emphasis on human capital, infrastructure, and human rights. The MDGs received limited interest from landscape architects.

A global foment about *starchitecture* and heritage collided in Vienna in 2005. As the International Federation of Landscape Architects (IFLA) Americas' representative, I spoke about defining urban landscape character and features, noting

> You must show up and speak effectively to contribute.

> Landscape architects have both embraced and disdained sustainability, with work that integrates these aspects or work that, obsessed with form and aesthetics, emphasizes alternate values and marginalizes our impact.

that half of city space was landscape (52 percent of Vienna, 56 percent of Washington, DC). My contributions and those of colleagues to this global dialogue led to the adoption of the UNESCO Recommendation on the Historic Urban Landscape (HUL) in 2011. HUL stressed that heritage and development are potentially mutually supportive, fostering social cohesion and quality of urban living. HUL tool groups (community engagement, knowledge and planning, regulatory systems, and finance) are being broadly applied, and landscape architects can benefit from partnering to activate them effectively.

A parallel movement recognizes intertwined culture and nature. As the 2015 IUCN/ICOMOS[1] Connecting Practices initiative and a growing body of literature have expressed, we must acknowledge the inseparability of culture and nature to survive. Humanity needs nature.

The millennium rumble has built to a roar with widespread opportunities to contribute. Embracing complex pervasive issues, the United Nations created, through an open two-year process, the United Nations Sustainable Development Goals (UN SDGs), adopted by 193 sovereign states on September 25, 2015. *Transforming our World: The 2030 Agenda for Sustainable Development* lays out 17 goals with 169 targets to guide nations toward patterns of development that favor diverse life on Earth. This transformational agenda is impressive because "never before have world leaders pledged common action and endeavor across such a broad and universal policy agenda." In parallel, the climate change forum at COP 21—the 2015 Paris Climate Conference—led to a promising global climate agreement.

Why should landscape architects engage in achieving the UN SDGs? Landscape architecture skills foster inclusive processes, partnering, and innovation that can rise to these massive challenges. The UN SDGs agreed on a worldwide framework, a complex global platform of goals and targets. If landscape architects take this framework seriously and partner to achieve the global 2030 agenda, we can make meaningful contributions to the momentum toward a sustainable future.

Which of the UN SDGs are relevant to landscape architects? A landscape architect might focus initially on Goal 13—Climate Action, Goal 14—Life Below Water, and Goal 15—Life on Land, as all are relevant. Digging deeper, the goals and specific targets reveal a multitude of ways to contribute individually with clients, teams, civic leaders, community partners, and collectively through professional organizations, leadership, and advocacy. For example, in Goal 1—No Poverty, the World Bank cites a veritable tidal wave of urban in-migration: five million people monthly seek an elusive better life, outpacing sustainable growth, increasing poverty, disparity, and inequality. Research and observation verify that few trees are found in disadvantaged neighborhoods. What can we do? Plant trees in poorer areas to build community wealth and advance social justice.

Goal 2—Zero Hunger is a key benchmark, as 800 million people go hungry every day. The targets address food security, nutrition, and sustainable agriculture with initiatives in urban agriculture, soil remediation, and market spaces as contributions. Goal 3—Good Health and Well-Being offers target 3.6 to "halve the number of global deaths and injuries for road traffic accidents." Landscape architects design better intersections, complete streets, and multimodal corridors. For Goal 4—Quality Education, we can serve as informants and advocates for sustainability and resilience. Targets 6.3, 6.5, and 6.6 of Goal 6—Clean Water and Sanitation address protecting water resources, counteracting pollution, and

Landscape architecture skills foster inclusive processes, partnering, and innovation that can rise to these massive challenges.

restoring water-related ecosystems, all things we already do. Goal 7—Affordable Clean Energy focuses on renewable energy, where landscape architects can contribute to siting and impact assessment. Goal 8—Decent Work and Economic Growth seeks to strategically "decouple economic growth from environmental degradation." For example, employment could spring from public spaces as we adapt and shape them anew.

Goal 11—Sustainable Cities and Communities addresses housing, transportation, planning, safeguarding cultural and natural heritage, provision of public space, and reducing disaster impact and environmental degradation. Target 11.7 seeks to "provide universal access to safe, inclusive, and accessible green and public spaces," engaging the urban commonwealth of public spaces. Accessible open spaces offer economic, environmental, and social benefits as cultural assets today and a legacy to the unborn. We can elevate public spaces.

Goal 16—Peace, Justice, and Strong Institutions and Goal 17—Partnerships for the Goals offer pivotal constructs for progress. In this urban century, recall Jane Jacobs' words that "cities have the capability of providing something for everybody, only because, and only when, they are created by everybody." Landscape architects must collaborate in shaping policy, planning, building, and setting new standards. The World Urban Campaign, Habitat III, and the 2030 Agenda await. Join me there.

Let's get to work, as President Barack Obama challenged us in his speech at COP 21, Paris, 2015

1.　The International Union for the Conservation of Nature (IUCN) and the International Council on Monuments and Sites (ICOMOS) published a report, "Connecting Practice," in 2015 that aims to explore and create new methods of recognition and support for the interconnected character of the natural, cultural, and social value of highly significant land and seascapes.

Patricia M. O'Donnell, landscape architect and planner, founded Heritage Landscapes LLC, Preservation Planners and Landscape Architects in 1987; the firm is dedicated to a vibrant future for communities and cultural landscapes. O'Donnell holds masters' degrees in landscape architecture and urban planning.

Chapter 26

LANDSCAPE ARCHITECTURE: NEW ADVENTURES AHEAD

DIRK SIJMONS

From my European perspective, the last 50 years have been marked by the growth and emancipation of our discipline. Landscape architecture has been freed from being the broccoli around the steak to becoming a discipline with its own agenda. The profession broadened its working field and its scope in that half century. It diversified. The traditional core of the discipline—making gardens and parks—expanded into public space design and the new field of retrofitting derelict industrial areas. The most talked about parks, like Parc de la Villette, Duisburg-Nord, and the High Line, are all brownfield parks. These parks have been pivotal in urban renewal processes and spurred real estate development.

For the discipline itself, however, the most important innovations were likely new ways of representing nature. For most of the modernist period, nature was reduced to a backdrop for leisure programs, but in recent parks representing nature has made a comeback. The formal architectonic representation in Parc de la Villette, the free range for spontaneous urban nature in Duisburg-Nord (where time itself seems to have become the main park theme), and the staged wild nature represented by configurations of Piet Oudolf's perennials on the High Line inspired by the self-seeded landscape that grew up between rail tracks—all show a distinct, more architectonic idiom being developed, even completed, by the "starchitects" of the profession.

Among the more anonymous professionals at the other end of the disciplinary spectrum, the expansion was almost as impressive. The traditional role of designing the landscape of infrastructure and agricultural development was steadily built upon to include designing new forests, water infrastructure, nature development, leisure projects, glass warehouse districts, airfields—in short, all the agents of change identifiable in the landscape that can be invited to make an expressive contribution to the landscape. In this century, urbanization itself was identified and addressed as a landscape architectural issue. Thomas Sieverts laid the analytical basis in his *Zwischenstadt* (Vieweg Braunschweig 1997) and that basis

To solve the current environmental crisis, we must retrofit our urban landscapes by patiently reweaving the extensive urban carpets into more sustainable configurations. We must improve the poorly functioning metabolism of these urban regions.

was instrumentalized by Charles Waldheim in *Landscape as Urbanism: A General Theory* (Princeton University Press 2016).

The 2014 International Architecture Biennale Rotterdam (IABR-2014) *Urban by Nature* took stock of the consequences. It positioned the viral growth of urban landscapes, especially the urban carpets in the large deltas of the world, in the wider context of the Anthropocene, the age of mankind.

The concept of the Anthropocene (Crutzen, 2000) produces some interesting side effects for our discipline. The once strict boundary between nature and human society is crumbling. Human interventions can be seen as, and compared with, forces of nature. We become aware that anthropogenic processes do not just influence geochemical cycles, biodiversity, sediment flows, ocean acidity, land use, sea use, and the climate, but that all these processes produce hybrids.

Anthropocene suggests that human and natural processes are linked in a complex whole. There is no initial situation or natural equilibrium to fall back on, just as we cannot go back to last week's weather. We can only go forward, and we must find the best ways of making progress and learning to ride this tiger.

The urban landscape is one of the largest of these hybrids. This artifact, formally known as the city, is our habitat. Most of the world's environmental problems have their roots in the urban landscape. To solve the current environmental crisis, we must retrofit our urban landscapes by patiently reweaving the extensive urban carpets into more sustainable configurations. We must improve the poorly functioning metabolism of these urban regions. This will involve projects focused on urban microclimate, sweet water supply, energy (transition), urban biogeography, disaster resilience, defragmenting the ecology, changing flows of sediments, and building material.

If we want to propel LAF's *1966 Declaration of Concern* another half century into the future and claim that our discipline is pivotal to solving these environmental problems, urban landscapes must be both the subject and the backdrop of our work. Landscape architecture is well positioned for the task.

First, we might be one of the few disciplines able to create a link between construction science and the life sciences. In this context, cultural representation of nature changes into mediating between nature and society. To my mind, without this vital mediation, we will find no lasting solutions for the problems of urban landscapes. Without it, the toolbox of the engineers and urbanists will be limited to half-baked technical solutions.

We might be one of the few disciplines able to create a link between construction science and the life sciences.

Furthermore, landscape architecture has a good track record in research-through-design with problems at the regional scale and at a landscape scale. In these (future) research trajectories, we learn to work in transdisciplinary teams where we will blossom as synthesizers, as generalists with a specific skill: (spatial) design. In addition to the natural sciences and the humanities, design is a third and separate domain of knowledge in which research uses modeling, pattern recognition, and synthesis. An appealing and unique element of research-through-design is that it can elegantly include the concept of free will. It concerns a future that is neither completely determined nor entirely unpredictable, but rather one that has, to some extent, a malleable character. Questions like "what can we want?" come with that territory. These plans have a new public. They are not aimed directly at execution but at influencing public opinion. Design can mediate between politics, other actors, and citizens. Design could even be instrumental to the design of the political will.

This type of work will require a stronger focus on understanding natural and social processes and design using their relationship with the spatial. Sometimes the scale will be geographic and raise questions: Is this still landscape architecture? Will landscape architecture preserve its disciplinary coherence with this broad scope and at these high levels of scale? Will the absence of real clients in these research-through-design trajectories loosen the ties with the programmatic precision? Is the connection with the genius of the place lost in a focus on processes? Or is the cultural element so homeopathically diluted that the lifeline to aesthetics, or even the discipline itself, snaps?

The overarching characteristic of all these scales and working fields is that specific link between program, place, and process. If this triad is present, we can still call it landscape architecture. A link with the spatial could function as a litmus test. Thus, projects in geoengineering or the acidity of the oceans would fail such a test.

But maybe we just need to stop worrying about these questions and restrict ourselves to the open-ended conclusion that landscape architecture is what landscape architects do.

Dirk Sijmons is a landscape architect and in 1990 cofounded H+N+S Landscape Architects. He studied architecture and environmental planning at the Technical University of Delft, where, since 2008, he has been professor of environmental design. Sijmons was curator of the International Architecture Biennale Rotterdam (IABR—2014) with the theme *Urban By Nature*.

Chapter 27

LESS A DECLARATION THAN SOME THOUGHTS

MARC TREIB

While I certainly have opinions, I am not really inclined to issue declarations. Although, as colonists, we signed one rather significant statement of that sort some years ago, in the end, declarations are not very American; we leave declarations and manifestos to the Europeans. To sign a declaration suggests a concordance of thoughts and values, a rare occurrence these days, or suggests having accepted the lowest common denominator. Today, at least regarding landscape architecture, that common denominator is stewardship and sustainable practice.

It is no secret that we live in a world of divisions and polarities, a world determined, sadly, by either/ors. We are divided by red and blue states, religions and nationalities, race, warring factions, and economic classes. Unfortunately, divisions also exist between those seeking absolute bio-atmospheric sustainability and those addressing the more aesthetic aspects of landscape architecture and, if one might use that troubled word, beauty. There is no argument that a framework for sustainable existence is very much needed and a prime concern for all of us. So, if we must make any declaration in this regard, we might call it the "Declaration of *Dependence*"—in this case, dependence on the planet, its atmosphere, its resources, and its natural systems.

But what of the concerns that lie beyond those of basic subsistence, those addressing the quality of life, human comfort, and even individual and collective pleasure? We do not love a place because it is sustainable. We love it for its qualities beyond those of performance. If the food tastes insipid, we do not care whether it is organic or not. Without aspirations

Without aspirations beyond achieving sustainability, the work of the landscape architect becomes only a form of environmental plumbing.

beyond achieving sustainability, the work of the landscape architect becomes only a form of environmental plumbing. We need plumbers, certainly, but we also need artists. The question, then, is what does the landscape architect contribute to the making of landscapes great and small that the biologist, hydrologist, or ethnographer does not? How does a grounding in the humanities as well as the sciences create a vision that contributes to more than mere environmental management?

An adage tells us that beauty is in the eye of the beholder. Yet, with only rare exception, cultural norms circumscribe, and therefore to some degree determine, personal aesthetic responses. Admittedly, no designer is capable of creating places perceived by everyone as beautiful or regarded as meaningful, even within the arena of a relatively small community.[1] We may have some agreement on functional issues and perhaps

on certain cultural values as well. But within any general consensus, considerable variations of opinion will inform the appreciation of beauty. Nevertheless, we can seek to create landscapes perceived as pleasurable and beautiful by a majority of those who visit or live in them, even if those efforts remain at the level of aspiration. Through an understanding of social mores, values, and tastes, the profession should be able to envision places exceeding in quality those already existing or those within the current memories of future occupants and users. Beauty is hardly the composition of form, space, and color alone, but more a conglomerate phenomenon that surpasses any single factor taken in isolation, just as true sustainability surpasses the checklist of factors needed to achieve LEED (Leadership in Energy and Environmental Design) certification.

Decades ago, in his own "gentle manifesto," the architect Robert Venturi claimed his embrace of "vitality as well

Through an understanding of social mores, values, and tastes, the profession should be able to envision places exceeding in quality those already existing or those within the current memories of future occupants and users.

validity" and "the difficult unity of inclusion rather than the easy unity of exclusion."[2] Can we not agree on an inclusive rather than an exclusive ambition for landscape architecture? Need we divide into separate camps those who stress the social, ecological, or aesthetic dimensions of landscape architecture considered only in isolation? The complexity of the commission may dictate where the stress in the design will fall; we know that time, money, and politics are often paramount in determining the design and the course of its realization. Or is that only an acceptable rationalization for dismissing concerns that also affect the lives of the individual and the community?

Sustainable is not antithetical to beautiful nor is beautiful antithetical to sustainable. Environments can, and I believe should, represent an approach of *both/and* rather than *either/or.* An exemplary landscape from the past is the Patio de los Naranjos in Seville, a setting that through almost a millennium of existence has provided a place of beauty, intelligence, and responsibility working together. Orange trees will not thrive in Seville's climate without human support. In creating the orange grove and its courtyard—originally the forecourt to a mosque—its makers devised a system of channels by which to irrigate the trees. They did not conceive their system of irrigation without a nod to beauty, however, and sought the exquisite as well as the functional. While providing a consistent water supply tempered by controlled evaporation, the courtyard's makers used the patterning of irrigation rills to invigorate the precisely executed ground plane of tawny brick. Standing apart from this tonal homogeneity are two overflowing fountains of marble that have shared water with the orange trees while standing as the focal points of the space.

With the Reconquest in medieval Spain, the Catholic Church superimposed a colossal cathedral upon the body of the mosque and in the process appropriated the Islamic courtyard and its orange trees. While an Islamic construction, this paradise of trees with their golden fruits possessed sufficient beauty to survive the conversion of ownership and religion. The courtyard was retained not for its harvest but for its beauty and symbolism. Beauty trumped the change of faith. While we appreciate aqueducts and expressways for their efficiency and performance, we also appreciate places

> Sustainable is not antithetical to beautiful nor is beautiful antithetical to sustainable. Environments can, and I believe should, represent an approach of *both/and* rather than *either/or.*

primarily or solely for their beauty, with a concord somehow collectively and perhaps even mysteriously achieved. There is no reason why a landscape cannot be sustainable, ecological, resilient, robust, enfranchising—or whatever the current buzzword or interest may be—and also be beautiful.

It seems to me that the challenge to landscape architecture today is not only to achieve peace with planetary systems but also to elevate pragmatic demands to the level of poetics—in other words, treat as poetry what may first appear to be prose. In his own declaration made some 40 years ago, the sculptor

Claes Oldenburg phrased it this way: "I am for an art that embroils itself with the everyday crap and still comes out on top."[3] There is a lot more crap today and more pressing issues than in the 1960s when Oldenburg made his statement, but also a lot more opportunity for pairing responsibility and beauty, not as an either/or but as a both/and.

1.   For four views of this thorny subject, see Marc Treib, ed., *Meaning in Landscape Architecture & Gardens* (Oxon, UK: Routledge, 2009)

2.   Robert Venturi, *Complexity and Contradiction in Architecture* (New York: Museum of Modern Art, 1966), 22–23.

3.   Claes Oldenburg, "I Am For," in *Claes Oldenburg*, eds. E. DeWilde and Alicia Legg (Amsterdam: Stedelijk Museum, 1970), 13.

Marc Treib is professor of architecture emeritus at the University of California, Berkeley, and a noted landscape and architectural historian and critic. He has published widely on modern and historical subjects in the United States, Japan, and Scandinavia.

Chapter 28

LANDSCAPE FUTURES

GERDO P. AQUINO

A person who never makes mistakes, never makes anything.
—Anonymous

As we ease into this century, our society is focused on reconnecting with ourselves, our ethnicity, our environment, and each other. In a traditional sense, our minds must be freed through the freedom of the body, and a park is a place where that can happen. A park is a place to unwind and reflect on things that matter most; a place for recreation to push the limits of the human body; and a place to understand nature and the spectrum of flora and fauna it supports.

Parks are at the core of what landscape architects do. The turn-of-the-twentieth-century Arcadian city park of green open lawns and axial cross paths has given way to twenty-first-century, highly designed, $20-million-an-acre, intensely programmed, public/private urban destinations, and we are just getting started. What will the twenty-second-century urban park look like? We can only imagine the possibilities.

Meanwhile, rapid advances in technology continue to deepen our understanding of the world by providing us with knowledge unimaginable just a decade ago: big data related to climate change and superstorms; drone-based high-resolution digital three-dimensional mapping of complex terrain to inform adaptive landscapes; and autonomous electric vehicles. Climatology experts predict that the recurrence of droughts in southwestern United States will continue into the foreseeable future with water distribution rights for agricultural industries, irrigation, and drinking water playing out in real time.

The landscape of the future will be challenging to imagine. The next generation of landscape architects must integrate evolving technologies into the context of the city park, generating interesting and unique outcomes.

As our cities continue to become more dense, we can expect even stricter regulations on our natural resources to balance the needs and wants of their communities. The landscape of the future will be challenging to imagine. The next generation of landscape architects must integrate evolving technologies into the context of the city park, generating interesting and unique outcomes.

Park design in the twenty-second century will lie at the nexus of four important considerations: mobility, ecology, programming, and financing. As independent considerations, these are simple mechanisms achieving a singular result. However, as a combined system moving at the same pace as our culture, these very mechanisms could yield real breakthroughs for park design in the twenty-second century and further reinforce the role of the landscape architect as the lead consultant overseeing large, complex public realm projects.

The work of landscape architects today will serve as the foundation for future innovation. I am inspired by the work carried out around the world by our peers in design and engineering, and the arts and sciences. I am inspired by civic leaders who understand the role of the landscape architect in helping make their cities more livable. A project like the Buffalo Bayou Watershed Park in Houston, designed by SWA, began as a grassroots movement centered around local flooding and open space issues and, over the course of 20 years, evolved into a national precedent for resilient landscapes addressing superstorm capacity/conveyance while providing a flexible open framework of community programming, regional bike trails, and habitats for birds and mammals. It is also now a top destination in Houston, helping elevate the identity of the city itself.

Park design in the twenty-second century will lie at the nexus of four important considerations: mobility, ecology, programming, and financing.

I am inspired by the renewed interest in activating the public realm in our great American cities. El Paso, Texas, recently redesigned its defunct 100-year-old public plaza into a vibrant, active hub of cultural programming and the arts. At the core of the design process was a highly visible group of constituents pushing for a strong narrative in the park's redesign, a narrative that would merge the past with the future while creating something entirely new—an emblem of the city's desire to step far into the twenty-first century.

I am inspired by cities like Belgrade, Serbia, which, over the course of four years, will transform a 1¼-mile stretch of postindustrial waterfront along the Sava River into an iconic open space, reconnecting its historic core with its waterfront.

We need all of you to continue thinking big, to explore the future of the landscape in a projective, experimental way, unhindered by preconceptions or rules, to take some risks and fail along the way, only to pick yourselves up and become stronger.

The new waterfront will also address rising flood levels in the Sava River by integrating a modular flood protection system activated by the hydraulic action of the rising water levels in the river. Projects like this bring much-needed open space to the dense urban core of Belgrade and provide residents and tourists with a safe, one-of-a-kind destination in Eastern Europe.

I am impressed by Cairo, Egypt, where they are contemplating moving the existing capital to an area east of the urban core to relieve congestion and provide a new urban fabric for millions of people. Fundamental to this move is a unique infrastructure strategy that would use both the Nile River and the Suez Canal to provide fresh, clean, sustainable water sources to the new development. At the center of this proposal is a landscape architect whose planning and design vision must be bold and visionary in the context of the storied history of Cairo.

Finally, I am impressed with the younger generation of landscape architects around the world. Wherever my travels take me, I inevitably meet bright young designers with big ideas and, more often than not, imbued with the kind of confidence and charisma it takes to change the world. We need all of you to continue thinking big, to explore the future of the landscape in a projective, experimental way, unhindered by preconceptions or rules, to take some risks and fail along the way, only to pick yourselves up and become stronger. Landscape architects are the future. Believe it.

Gerdo P. Aquino, is firmwide chief executive officer of SWA, a landscape architecture, urban design, and planning practice. Aquino is also adjunct associate professor at the University of Southern California. He holds a bachelor of science in landscape architecture from the University of Florida and a master of landscape architecture from Harvard Graduate School of Design.

PART V

VOICES OF THE FUTURE

AS LANDSCAPE ARCHITECTS, WE VOW TO CREATE PLACES THAT SERVE THE

HIGHER PURPOSE OF SOCIAL AND ECOLOGICAL JUSTICE FOR ALL PEOPLES

AND ALL SPECIES. WE VOW TO CREATE PLACES THAT NOURISH OUR DEEPEST

NEEDS FOR COMMUNION WITH THE NATURAL WORLD AND WITH ONE ANOTHER.

WE VOW TO **SERVE THE HEALTH AND WELL-BEING OF ALL COMMUNITIES**.

FROM THE NEW LANDSCAPE DECLARATION

Chapter 29

A CALL FOR BROADENED COMMUNICATIONS AND CRAFT

JOANNA KARAMAN

The *1966 Declaration of Concern* was aimed at shaping the next generation of practitioners. In reaction to this manifesto 50 years later, the discipline seeks to reevaluate its position and value in contemporary society. The way we as landscape architects communicate our skills and design intent will ultimately dictate our work done in practice.

The declaration asserts that "there is no 'single solution' but groups of solutions carefully related one to another." This call for "collaborative solutions" is the key to the future of landscape architecture. Suspended between science and art, landscape architects are trained to synthesize information and coordinate many different competing interests. However,

the first step toward using those skills, even before the design process begins, is to make sure that we have a seat at the decision-making table. The perceived value and understanding of our work is something that is still being defined by the relatively young profession and presents itself as an opportunity to work in a meaningful way in the future.

Landscape architects cannot be expected to solve the socio-environmental crisis of the twenty-first century single-handedly; we thrive by working with collaborators who share our values. Further, we must still enjoy our craft and explore it playfully through a range of mediums and expressions. Experimentation and creation are at the core of our practice and distinguish landscape architects from environmentalists and policy makers. And finally, if not foremost, we must establish an economic independence that will allow us to pursue projects and other endeavors proactively as opposed to being trapped at a receiving end of client demands. Our

The perceived value and understanding of our work is something that is still being defined by the relatively young profession and presents itself as an opportunity to work in a meaningful way in the future.

training positions us to lead large-scale urban projects and initiatives with landscape intelligence, which is why being involved at the start of, or before, design is crucial.

While the potential of landscape architecture is great, we as practitioners need to have a willing public, a captive audience. Furthermore, the pleasure and craft of our work need not be in opposition to our role as advocates.

"Does our pain for the world, our knowledge of apocalypses to come require us to forgo pleasure altogether?…So there we landscape architects stand, perched on a wobbly crate in the town square, delivering our lugubrious sermon to an indifferent world as it rushes past to the Apple Store."[1]

Although a comedic hyperbole, this quote from Lickwar and Oles encourages us to rethink our strategy of creating public awareness. The technological advances made in the past 50 years should not be shunned or ignored but rather embraced as a powerful toolkit that we can deploy—not

conform to—in both communication and practice. Whether in three-dimensional modeling or time-based media (film and animation), the way we begin to represent landscapes can more accurately describe the dynamic environments of proposed designs.

Further, accessibility, as opposed to efficiency, is the real power of today's new methods of work. For instance, the general public may not understand a technical section or plan, but a short film can quickly be understood by all, convey emotion, and even help spark new ideas. People value what they can see and understand. In turn, this social value translates into cultural and economic agency for the practice of landscape architecture. A poignant piece of writing, graphics, or time-based media could help to change public perception more than an oversimplified "greening" project that presents a false sense of security about the state of the environment. We need to be honest about the way we represent and ultimately build our future landscapes.

Landscape architects have opportunities to take on representation beyond the static image. Rather than an afterthought to design, contemporary media can help shape our design and research processes and create better links to the place that the built environment holds in the minds of the public today.

"The influence of contemporary film and communications media on landscape appreciation has yet to be fully studied…. It is immense, especially in American popular culture. These effects go both ways, of course, for the building of new landscapes and their subsequent representation in art can also affect the evolution, value, and meaning of larger landscape

As landscape architects, we need to find a way to operate from both ends—working on shifting the public will toward a true change in response to environmental pressure, as well as continuing to create and construct meaningful places that add social and ecological value.

ideas as well as other cultural practices….The reciprocal interactions between the built and the imaginary lie at the center of landscape architecture's creativity."[2]

Hence, a better understanding of the relationship between the built environment and media can augment the framework in which we as landscape architects work, as well as enforce our role as ecological stewards in the public realm. The permanence and legacy of our public projects is as much determined by societal value as the actual durability of materials. By instilling a sense of stewardship in the people who these public spaces serve, we can help preserve urban landscapes that already exist, as well as help to promote the creation of new ones.

To be leaders in the coming decades, we need to leverage both our skills and our projects in the public realm, changing the way we work and the way we are perceived. This shift in the status-quo paradigm is crucial if we hope to implement meaningful projects in this age of environmental crisis.[3] The final point in the program put forth by the *1966 Declaration of Concern* calls for "a nationwide system for communicating the results of research, example, and good practice." In order to build off of this, we must spread this information beyond our discipline, beyond the context of the New Landscape Summit, in a creative and effective manner. Hence, as landscape architects, we need to find a way to operate from both ends—working on shifting the public will toward a true change in response to environmental pressure, as well as continuing to create and construct meaningful places that add social and ecological value.

1.   Phoebe Lickwar and Thoams Oles, "Why so Serious, Landscape Architect?," *LA+* 1, vol. 2 (2015): 82-83.

2.   James Corner, "Introduction: Recovering Landscape as Critical Cultural Practice" in *Recovering Landscape: Essays in Contemporary Landscape Architecture* (New York: Princeton Architectural Press, 1999), 9-10.

3.   Donella H. Meadows and Diana Wright, *Thinking in Systems: A Primer* (White River Junction, Vt.: Chelsea Green Pub., 2008). Meadows and Wright's book outlines a series of methods and "leverage points" by which to enact change in a system. One of the most powerful tools they list is to change the current paradigm of society.

Joanna Karaman is a Los Angeles-based landscape designer at OLIN. She graduated in 2015 from the dual-degree master of landscape architecture and master of architecture program at the University of Pennsylvania. She was the 2015 Olmsted Scholar for the University of Pennsylvania.

Chapter 30

EXPERIMENTAL LANDSCAPES: THE POWER OF THE PROTOTYPE

NINA CHASE

We are in the midst of a refreshing paradigm shift. Pop-up parks, guerilla wayfinding, temporary parklets, DIY bike lanes—you cannot scroll through an article about urbanism without finding projects that are framing the process of city building as participatory, fun, and chock-full of experimentation. Designers, activists, and even developers are heeding the call, taking to the streets, quite literally, to prototype the potential of their cities' public realms. Most notably, in New York City, Times Square has experienced a pedestrianized transformation that began with a tactical takeover by the New York City Department of Transportation putting out lawn chairs.

Conceived and implemented as pilot initiatives, short-term projects position landscape solutions as first-phase building blocks for cities and have found a home within the tactical urbanism movement. Sensory and interactive, this creative approach invites stakeholders to connect with a future investment, albeit at a smaller scale and with less financial risk. Tactical urbanism projects show citizens what they are buying before they write the check. Phasing, flexibility, evolution—words once confined to the playbook of landscape urbanists are making their way into common practice. But the tenets of tactical urbanism—quicker, lighter, cheaper—should not be limited to temporary solutions. The movement has the potential to tackle larger environmental challenges and be led by the original urban placemakers: landscape architects.

Indeed, cities are resolving some of their most pressing challenges by using landscape prototypes to demonstrate and inspire systemic change. In Raleigh, a guerilla signage campaign aimed at encouraging people to walk resulted in a

> Conceived and implemented as pilot initiatives, short-term projects position landscape solutions as first-phase building blocks for cities and have found a home within the tactical urbanism movement.

new pedestrian plan for the city. In San Francisco, Park(ing) Day became an international phenomenon that has since led to the transformation of parking spots into permanent parklets. In New York City, Janette Sadik-Khan, a former commissioner of the city's Department of Transportation, wielded pilot testing as her secret weapon to transform city streets. "Instead of arguing and debating, try something first and give people something to experience," she said in a recent interview. Her proof-of-concept pilots included painted bike lanes and lawn chairs in Times Square. The efforts led to the transformation of more than 400 miles of streets, integrating bike lanes, safer pedestrian crossings, and narrowed vehicular lanes.

More and more, cities are searching for solutions to large-scale environmental and societal pressures. Tactical urbanism's can-do, optimistic approach has the potential to galvanize support for long-term landscape strategies well beyond the scale of the parking spot. Sea-level rise, underused vacant land, and long-term droughts are issues on which citizens and municipalities often have difficulty finding common ground. To complicate matters, these issues are occurring at the scale of whole regions, which often lack a single entity that can act unilaterally to address them.

The profession of landscape architecture is equipped with the skills to untangle these thorny problems by coupling the efficiency of engineered solutions with the sensitivity of cultural and ecological opportunities. But because it is difficult to show what nature-based projects might look like until they are finished in 5–30 years, garnering support is tough. It is not easy for the average citizen to imagine what a resilient coast could look like or what new environmentally focused land

> Tactical urbanism's can-do, optimistic approach has the potential to galvanize support for long-term landscape strategies, well beyond the scale of the parking spot.

uses will mean for their neighborhoods. Long-range landscape planning takes time, political will, money, and community patience. Pilot landscape projects that embody the "test before you invest" mantra could help synthesize collective visions toward a landscape-centric future.

Consider sea-level rise as one example. Scientists project that tides will rise two feet by midcentury and six feet by 2100. This new tide line will transform the world's coastal landscapes. Designers have posited optimistic visions for our new wet future. Concept renderings show raised sidewalks, floating buildings, and protective wetlands. But ask the average citizen what it means to build a resilient coast and you will often be met with blank stares. As coastal cities plan for long-term investments, pilot projects could demonstrate the potential of resilient landscape solutions sooner rather than later.

Imagine a segment of an existing seawall in South Beach in the city of Miami Beach transformed to include seating, animal habitats, or recreational features (e.g., a climbing wall at low tide). Today's seawalls serve only one purpose—to keep the water out—but given the lack of valuable space in our city, coupled with the need for better protective measures, our coastal edges should serve more than one role. Cities could use pilot projects to test creative ideas for multipurpose seawalls before investing millions in storm surge infrastructure. Existing seawalls would provide necessary protection, while a series of newly designed temporary facades could illuminate the possibilities for multifunctional projects.

As another example, envision the Boston Harbor Islands being transformed into a testing ground for a variety of resilient coastal strategies. An island could become a research hub for experimenting with new edge conditions, where designers, ecologists, and engineers could document and quantify the protective effects of dunes, saltwater marshes, multipurpose seawalls, or small floodgates. The tactical takeover of one or more islands could demonstrate the potential of innovative shoreline structures and landscapes that could then be applied across the city.

Cities are facing environmental and social threats that will affect urban development for the next several decades. As a planning and design tool, tactical urbanism has proven effective

Long-range landscape planning takes time, political will, money, and community patience. Pilot landscape projects that embody the "test before you invest" mantra could help synthesize collective visions toward a landscape-centric future.

in generating short-term action around long-term change, but it does not absolve those at the helm of city planning and design from committing to planning rigor, political process, and public investment. Where traditional planning approaches—so often opaque and abstract—fail to ignite public passions, tactical urbanism can add a dose of accessibility, whimsy, and experimental fun. Designers, city officials, community members can engage with prototypes, collectively iterating and creating a dialogue around what works and what does not. Cities are increasingly positioning innovative public spaces and nature-based solutions as necessary to sustainable urban growth. The profession of landscape architecture is poised to claim short-term projects as harbingers to long-term urban transformations. By harnessing the power of the prototype, we can show, not just tell, what is possible for our cities' and our profession's futures.

Nina Chase is senior project manager at Riverlife in Pittsburgh. Previously she was an associate in Sasaki Associates' Urban Studio in Boston. Chase earned her master of landscape architecture from Harvard Graduate School of Design and bachelor of science in landscape architecture from West Virginia University. Chase was the West Virginia University Olmsted Scholar in 2009.

Chapter 31

DESIGNING CONSTRUCTED ECOSYSTEMS FOR A RESILIENT FUTURE

SARAH PRIMEAU

As landscape architects, we must become deliberate and skillful designers of constructed ecosystems. For too long we have manipulated ecosystems, altering the flows and cycles of water, materials, and species without adequate ecological knowledge, and without understanding the broader harm or benefits of our actions. We have designed landscapes to provide short-term benefits to the human species only, forgetting that our ultimate well-being and quality of life depend on resilient, functional ecosystems.

In the same way that we have adopted low-impact development practices and native planting, we must now embrace a rigorous new approach to ecological design,

particularly in the context of constructed ecosystems. Landscape architects are well positioned to lead the design of constructed ecosystems in this century due to our expertise in working across disciplines and scales, and I believe our role will be strengthened by embracing the following propositions.

Designing for ecosystem services

In the current era of rapid global change, we must learn how to design landscapes for ecosystem services, creating regenerative natural systems that produce clean water and fertile soils and that support a rich web of species, including humans. These services are being rapidly lost due to accelerated climate change and conventional patterns of development.

We must also be mindful that designing ecosystems is inherently political, and one with important ethical

In the same way that we have adopted low-impact development practices and native planting, we must now embrace a rigorous new approach to ecological design, particularly in the context of constructed ecosystems.

considerations. Choosing which services we want ecosystems to provide, and to whom, is a subjective decision. While our role in manipulating ecosystems is not new—after all, humans have hunted large predators to extinction, cultivated shellfish, and changed plant communities using fire for millennia—we need to recognize a moral responsibility to maintain the production of ecological services that may not be of direct benefit to us personally. By definition, ecosystem services benefit humans, but they also support the more than nine million other species with which we share the planet.

Adapting to climate change

We must work more strategically across scales and move beyond abstract ideas of ecology in order to help human and nonhuman communities adapt to our changing climate.

Advocating for the protection and reestablishment of regional ecological networks will be increasingly important to support species migration and adaptation. We simultaneously need to work with planners and ecologists to translate regional priorities into site specific, on-the-ground design decisions. Constructed ecosystems within these networks can support biodiversity and contribute to the resilience of larger ecosystems.

Landscape architects will also play an increasingly critical role in an era of temperature and precipitation extremes by helping to moderate climatic effects through local microclimate improvements.

Our creative problem-solving skill set will also be important in helping coastal areas adapt to sea-level rise. Instead of implementing single-function solutions such as seawalls and dikes, we must encourage adaptation in a way that regenerates productive intertidal ecosystems and increases the resilience of coastal communities—both human and nonhuman—over the long term.

Finding synergies with engineered ecological processes

Within built environments, our greatest opportunity for creating constructed ecosystems is by finding synergies with engineered ecological processes, such as stormwater management and wastewater treatment systems. Urban infrastructure is still largely based on single-purpose hard engineering solutions designed to serve functions formerly provided by our ecosystems. Moving from single-purpose gray infrastructure to multipurpose green infrastructure provides us with the opportunity to design constructed ecosystems that provide a broader suite of ecosystem services. New or restored urban aquatic ecosystems, for example, can be constructed to capture, filter, and convey rainwater, while recharging groundwater, creating ecological corridors, moderating temperature extremes, and creating inviting places for people to enjoy.

Reconnecting people with nature and fostering ecological literacy

Increasing the resilience of our societies also requires us to design our cities in a way that supports a meaningful connection between people and nature. The steady migration of people into cities requires us to bring nature into the cities as well.

Giving people the opportunity to experience natural phenomena in their daily lives—such as fish spawning in the city or beavers building a dam in an urban wetland—can foster ecological literacy by helping build an understanding of our interconnected world.

Improving people's access to designed ecosystems can also help cultivate an environmental stewardship ethic by increasing people's understanding of the need to conserve and regenerate ecosystems outside of city boundaries.

These experiences also help build people's understanding of place—of where they are situated within the world, how that landscape was formed, what plants and animals it supported, which people have depended on this land, and what role it played in the development of our cities.

Embracing evidence-based planning and design

Accomplishing these ambitions requires us to improve our understanding of ecosystem processes and ecosystem development over short- and long-term time frames. Increasing our collaboration with ecologists can help us translate scientific evidence into tangible design principles and can help

Together with allied professionals in planning, ecology, engineering, and public health, among others, we are well equipped to synthesize diverse ideas from a variety of fields and translate complex regional goals into tangible on-the-ground design responses.

us set priorities for new ecological research. Basing our plans and designs on the best available evidence enables us to build on lessons learned and use limited resources effectively.

Being ambitious

While some of these ideas are not new to our profession, we must be more ambitious in advocating for their implementation. We need to position ourselves to lead multidisciplinary design teams and to extend our influence across all levels of government.

Together with allied professionals in planning, ecology, engineering, and public health, among others, we are well equipped to synthesize diverse ideas from a variety of fields and translate complex regional goals into tangible on-the-ground design responses. Landscape architects are uniquely suited to achieving these ambitions as we embody a profession that has, at its foundation, the objective of articulating hopeful visions for the future.

Sarah Primeau is a landscape architect at space2place design in Vancouver. She holds a master of landscape architecture from the University of British Columbia and a master's in wetland ecology from the University of Waterloo. Primeau was an Olmsted Scholar Finalist in 2011.

Chapter 32

THE REPARATION OF OUR EPISTEMIC RIFT AND A RETURN TO VALUES

SCOTT IRVINE

Landscape architecture has become an urban discipline. The profession is being focused on the social and therefore, in socially dense places, prioritizing direct daily experience with design interventions that see immediate benefits and impacts.

Nonurban landscapes and places of wildness are becoming detached from landscape architecture amid intensifying urbanism, and projects that will be directly experienced by few, if any, people are increasingly difficult to justify. However, these places have a disproportionate importance to us. Despite being outside of our daily experience and often being places that require a concerted effort to immerse ourselves in, many of our personal, collective, and national values lie outside of cities and are predicated upon the existence of wilderness.

With 80 percent of Canadians and Americans living in urban areas, it is clear why landscape architecture has shifted its focus. However, with urban land coverages of just 0.25 percent and 3 percent respectively, each country has over 9.5 million square kilometers of nonurban land that have been losing focus in the discipline.

This urban shift is about much more than the projects that are undertaken; contemporary landscape architecture has developed primarily through the lens of cities. Knowledge, ethics, and modes of operating are now reified in an urban environment.

The unprecedented nature of urban environments has continually narrowed the scope of regionalism to the borders of cities themselves. There is now a push to solve all the problems caused by large-scale urbanization within the cities themselves, rather than considering a broader array of strategies and better understanding of how cities are situated, as places and as Earth systems,[1] in a much broader regional context.

Amid the shift toward social stewardship, I believe that landscape architecture needs to reinstate landscape stewardship. Although ecological work has indeed been

> There is now a push to solve all the problems caused by large-scale urbanization within the cities themselves, rather than considering a broader array of strategies, and better understanding of how cities are situated, as places and as Earth systems, in a much broader regional context.

undertaken by landscape architects in urban environments, it is primarily socially driven. We ought to be able to take the position that wilderness has a right unto itself to exist and consider that we might have ethical responsibilities to landscape processes and Earth systems themselves.

Many concepts fundamental to how we have situated ourselves with respect to wilderness and the nonurban are now understood to be false. Ecological models often actively exclude human influence or attempt to measure it with the aim of reversing it or offsetting it in equal measure. Much has come to light that strongly suggests that the landscapes, the wildernesses, of North America were highly managed systems before European modes of operating in the land became predominant. We are at a crossroads, I believe, because our knowledge base is being developed specifically in response to urban environments. The discipline has not been working toward driving new knowledge, ethics, and modes of operating in wilderness and nonurban environments.

Jeremy Vetter uses the term *epistemic rift* to describe the divide we place between humans and nature. Although there is no actual separation between ourselves and nature, we have developed fundamentally different ways of understanding each. The reparation of this is as much about resituating ourselves in wilderness and in nonurban landscapes as it is about reconceptualizing urban environments themselves. The latter is what has been driving contemporary landscape architecture; however, I think the former has been left behind.

As much as the Anthropocene is conceived by the spatial scale of our actions in the land, it is more a profound subversion of the temporal scale of a given process. The spatial consequences are evident, but more dramatic is the way we now drive the time scales of Earth systems. Coming to terms with the Anthropocene offers an opportunity to consider not only how we situate ourselves in the world but also how we situate ourselves in time. We might shift toward establishing values that respond to the immediate human time scale but also transcend generations—to create what J. B. Jackson calls "a sense of place, a sense of time."

Landscape architecture needs to actualize itself within this ongoing epoch; the Anthropocene is not a condition we should simply respond to, it is something that we continually create. Further, it is not an epoch that we can design ourselves out of. This is an opportunity to make a fundamental shift from landscape architecture as reparation to a discipline taking active agency in creating places we aspire to.

Frederick Law Olmsted's approach to landscape architecture was very much about the creation of values. Although his work is now understood primarily within urban environments, it was not created through the lens of the city. I think that the discipline might reprise itself as a fundamental generator of values in the landscape, especially outside of the urban environment, in the kinds of places where some of our most important personal and collective values are held.

The discipline of landscape architecture ought to maintain its breadth across the spectrum of environments—from wilderness to the rural and to the urban. This spectrum of environments is highly connected, both in terms of Earth systems and in the way our individual and collective identities are formed. In the rush to positively affect many people directly, the opportunity to affect many more in much more profound and elemental ways has languished.

A continual reevaluation of the values that underpin any given project is fundamentally a good thing, and this could

> Landscape architecture ought to operate more holistically in terms of what it accomplishes as a discipline and leave room for the exploration of what it is we can accomplish.

be done mindfully with an eye toward better understanding the potential of the discipline. Landscape architecture ought to operate more holistically in terms of what it accomplishes as a discipline and leave room for the exploration of what it is we can accomplish. I think that the presence of outliers, operating at the fringes of landscape architecture, will always be a good thing.

1.   Earth systems is defined by Clive Hamilton as the science of the whole Earth as a complex system beyond the sum of its parts, distinct from the environment.

Scott Irvine earned a bachelor of environmental design in the landscape and urbanism option and is currently a graduate student in the master of landscape architecture program at the University of Manitoba. His first experience in the profession came at Basterfield & Associates Landscape Architects in Peterborough, Ontario. Irvine was the University of Manitoba Olmsted Scholar nominee in 2015.

Chapter 33

LEADERSHIP THROUGH LISTENING

TIM MOLLETTE-PARKS

It is getting noisy out there.

In the 50 years since the *1966 Declaration of Concern*, new technology has emerged and engendered a cacophony of voices about our world. Good ideas abound more than ever before, but the multiple ways of communicating are constantly shuffling the deck; the good ideas are hidden among pretenders. In this new reality, listening is the challenge.

The process of landscape architects designing public space mirrors this societal trend in many ways. Even in my short career span, I am stunned by the range of voices I have worked with in developing landscape designs—museum curators in Los Angeles and New York; mayors in Charleston, South Carolina, and Oakland, California; environmental activists in Jackson, Wyoming, and Memphis, Tennessee; civil rights activists in the Watts neighborhood of Los Angeles; software

engineers in Silicon Valley; children of all ages and their teachers, mothers, fathers, brothers, sisters, and friends. And from all of their points of view—intensely disparate as they are—the landscape has something to offer all of them. And they are acutely aware of this.

This range of voices has had significant influence on design practice. In too many cases, strategies for dealing with this influence have turned toward hyperspecialization and overdependence on personality-dominated work (*starchitecture*, if you will). These are two ways of working that, frankly, do not value listening. The result is often work that is formulaic or self-referential—two things the landscape must never become.

So, how can landscape architects ensure that our work transcends the growing shrillness of public discourse and intense variation of user input without becoming formulaic or self-referential? My personal approach requires leveraging two aspects of landscape architecture that are not really new but with which we desperately need to reconnect.

> We learn to observe patterns of drainage, vegetation, fauna, wind, and sun. We learn to watch people move through, occupy, and manipulate a landscape. We observe these patterns not just for their potential performance but, just as importantly, for their poetry.

First, we are listeners. From the beginning of our training, we are taught to listen to and read a landscape. We learn to observe patterns of drainage, vegetation, fauna, wind, and sun. We learn to watch people move through, occupy, and manipulate a landscape. We observe these patterns not just for their potential performance but, just as importantly, for their poetry.

Second, we are designers. Our legacy is Frederick Law Olmsted, André le Nôtre, Gertrude Jekyll, Thomas Church, Garrett Eckbo, Dan Kiley, Peter Walker, Lawrence Halprin, Laurie Olin, and Ed Haag. These practitioners understood form, pattern, and texture. They understood scale, proportion, and tectonics. We cannot lose these. We cannot exchange them for complex three-dimensional models and eye-popping renderings and pretend they are somehow the same thing.

> It is getting noisy out there. We know how to listen. Now should be our time to lead by listening.

On some occasions, our profession has relinquished design thinking to our architecture counterparts, who may have the best of intentions but likely have not had our training in listening to a landscape. We cannot do that anymore. The role of the landscape and the landscape architect is too important.

Of course, we need to continue our focus and drive toward sustainability and ecological performance. We should continue to cultivate our techniques and refine our methods for this portion of our work. But sustainability and performance need to be the base, not the ultimate characterization of our work. We are past that.

We must be able to devise clear and delightful design gestures in the landscape just as easily as we size a swale. Why? Because the spaces that endure are the ones that are imbued with too much meaning and too much beauty for the descending generations to remove them. That defines sustainability and performance over time.

Lucky for us, we have partners to help unlock the beauty and enduring nature of a project. We have children and software engineers and mayors and civil rights activists. They love the landscapes they live in. We can lead them by listening. And during the listening, we must apply the environmentally restorative approaches we care so much about along with the poetry of design (the *freedom of musicians,* to paraphrase Jens Jensen's *Siftings*). This will not only distinguish us within the design and engineering professions, it will differentiate us among modern society.

It is getting noisy out there. We know how to listen. Now should be our time to lead by listening.

Tim Mollette-Parks is associate principal at Mithun in Seattle. He earned his master of landscape architecture at the University of California, Berkeley, where he has also served as guest instructor the past four years, teaching design courses in the Department of Landscape Architecture and Environmental Planning. Mollette-Parks was a 2009 National Olmsted Scholar Finalist.

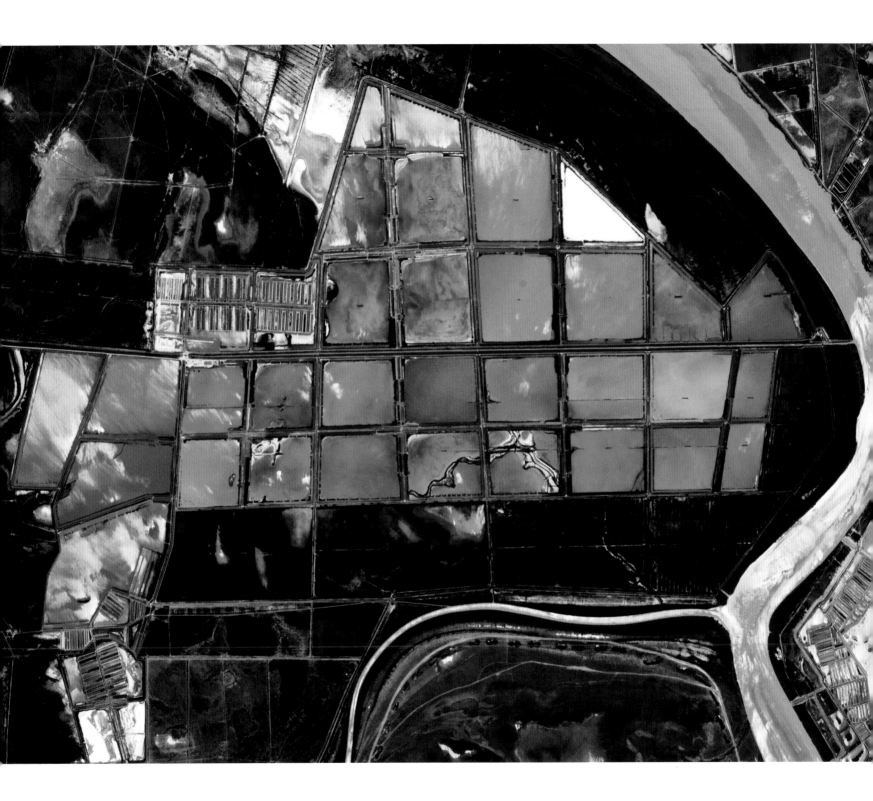

Chapter 34

LANDSCAPE AS NARRATIVE

AZZURRA COX

Before he became known for his nuanced understanding of ecological dynamics, Frederick Law Olmsted chronicled the social dynamics of a young nation. As a journalist, he confronted a uniquely fraught period in the American South on the brink of the Civil War. At the beginnings of westward expansion, he wrote about life on the frontier and advocated for the national park movement. That singular spirit of *communitiveness* that would later propel his landscape projects was rooted in the narratives—intimate and grand, personal and societal, spanning years and decades and, indeed, nearly a century—that he had worked tirelessly to reveal and understand.

Olmsted's vision for landscape as a powerful political construct was ahead of its time, and his commitment to creating new kinds of public spaces, steeped in complexity and imbued with agency, is a legacy my generation of landscape architects must embrace. As we look to the next 50 years and craft our roles as spatial and relational thinkers, our task remains to first reveal and understand the sociopolitical dynamics that shape our cities, our territories, and our social imaginaries (set of values). In order to do this, we must be willing to step outside of our disciplinary boundaries—to question and observe and listen deeply, to leave behind the guise of expert, to seek collaboration, and to always welcome the role of student.

If a deep engagement with physical and social realities becomes the unified base for our disciplinary future, the question of where and how we, as individual practitioners, choose to engage remains as open and multifarious as the world around us. Indeed, at the core of this declaration is a call for a new multiplicity of intentions, agendas, and commitments. This is a call for challenging and expanding the narratives we, as design students and practitioners, consider part of the conversation—with the aim of not only addressing present questions, but also shaping future answers.

At its core, design is a creative, narrative act. With each design decision, landscape architects can shape, heighten, or ignore the narratives that live and breathe in a site. In doing so, they speak to evolving identities of space and place. When we

We must be willing to step outside of our disciplinary boundaries—to question and observe and listen deeply, to leave behind the guise of expert, to seek collaboration, and to always welcome the role of student.

design, whose histories are we revealing and what futures are we projecting? What are the tensions between the designer, the changing, breathing material of landscape, and the network of stakeholders? In their narrative choices, landscape architects bear significant responsibility toward crafting experience, fostering cultural and environmental stewardship, and evoking something simple yet radical: meaningful engagement with a place and fellow citizens.

My own development as a designer is grounded in the conviction that all built space is inherently political and as such, landscape architecture is a political discipline. Landscape architecture is also a key medium through which to draw connections between environmental and social dynamism. At a time when so many scales of conflict—from climate change and resource scarcity to urban segregation

and police brutality—are intricately tied to intentional spatial systems of organization, landscape architecture can become a unique tool for change thanks to this ability to connect on both physical and metaphysical levels.

Drawing from ecological principles, the concept of fragmentation is perhaps useful in defining current relationships between the environment and society—and outlining some of the many possible future grounds for the discipline. When we speak of urban fragmentation in postindustrial American cities, for example, we allude to dynamics at multiple scales of experience. The same systematic divestment policies that have led to cities like St. Louis being more segregated than ever—and that speak clearly to social fragmentation along lines of race and class—have also resulted in the breakdown of ecological connectivity. The contested public spaces that form the fabric of a city like Detroit— or rather, its degradation over time—are the embodiment of that breakdown. The role of the

With each design decision, the landscape architect can shape, heighten, or ignore the narratives that live and breathe in a site.

landscape architect there might be to design spaces where the value of ecological recovery is deeply related to a commitment to social investment and a stand against systemic violence of any kind. I am considering the potential of this role in the case of St. Louis's Greenwood Cemetery, an African-American cultural landscape that has suffered the effects of divestment and prompts deep questions about remembrance and erasure as embodied in the land.

When considering territorial-scale challenges, too, fragmentation provides a useful conceptual framework for the landscape architect. Last summer I researched the traditional and changing roles of water across the Bolivian altiplano, linking the centuries-old aquatic landscapes of Lake Titicaca with the emergent urban rivers of El Alto. As part of one watershed, the two conditions are innately related, yet their fragmentation in the collective imaginary has led to a rupture, and urbanization behaviors in El Alto are directly degrading Lake Titicaca's delicate ecosystems. There, reimagining the riverfront as public space might both begin to restore ecological balances and recover those less tangible connections.

Yet, we must be more than problem solvers. By crafting sensorially memorable experiences, we must help generate new modes of living and ask new questions. How might spatial quality engender interaction? How might the unexpected promote openness? How might poetry foster stewardship? Above all, we must remember that landscape is uniquely telescopic: it has an uncanny ability to reference memory and the future, the material, and the abstract. As Robert Smithson observed of Olmsted's work in "Frederick Law Olmsted and the Dialectical Landscape" (Artforum 1973), the most powerful landscape projects are "never finished; they remain carriers of the unexpected and of contradiction on all levels of human activity, be it social, political, or natural."

Our task for this century is to craft those vessels of human experience and agency, to balance between the extreme specificity of a site and an openness of vision that welcomes a range of voices, subjectivities, and tensions. Designing space is a necessarily humanistic endeavor; it is messy. It asks us to both speak to current social realities and ignite radical new ones. Therein lies our greatest potential.

Azzurra Cox is a landscape architect at GGN in Seattle. She graduated with a master of landscape architecture from Harvard Graduate School of Design and a bachelor of arts in social studies from Harvard College. Previously she worked as an editor at The New Press. Cox was a 2016 National Olmsted Scholar.

PART VI

THE CALL TO ACTION

TO FULFILL THESE PROMISES, WE WILL WORK TO STRENGTHEN AND DIVERSIFY OUR GLOBAL CAPACITY AS A PROFESSION. WE WILL WORK TO CULTIVATE A BOLD CULTURE OF INCLUSIVE LEADERSHIP, ADVOCACY, AND ACTIVISM IN OUR RANKS. WE WILL WORK TO RAISE AWARENESS OF LANDSCAPE ARCHITECTURE'S VITAL CONTRIBUTION. WE WILL WORK TO SUPPORT RESEARCH AND CHAMPION NEW PRACTICES THAT RESULT IN DESIGN INNOVATION AND POLICY TRANSFORMATION. **WE PLEDGE OUR SERVICES. WE SEEK COMMITMENT AND ACTION FROM THOSE WHO SHARE OUR CONCERN.**

FROM THE NEW LANDSCAPE DECLARATION

Chapter 35

AESTHETICS PANEL

PANELISTS

Claude Cormier, Principal, Claude Cormier + Associés

Maria Goula, PhD, Associate Professor, Department of Landscape Architecture, Cornell University

Mikyoung Kim, Founder and Design Director, Mikyoung Kim Design; Professor Emerita, Rhode Island School of Design

Chris Reed, Founding Principal, Stoss; Associate Professor in Practice of Landscape Architecture, Harvard Graduate School of Design

Ken Smith, Principal, Ken Smith Workshop

MODERATOR

Adam Greenspan, Design Partner, PWP Landscape Architecture

The Aesthetics panel was charged with assessing the status of aesthetics and beauty in the profession and its place in the new declaration. The following is a condensed and edited version of the panelists' remarks and is not intended to be a verbatim or complete representation.

Adam Greenspan: There seems to be a deep-seated fear—or at least a warning—that if you acknowledge aesthetics or beauty directly, your work or the field will be marginalized. When we are thinking about a declaration for the future, it should be acknowledged that the physical designed and constructed landscapes that we create engage nature, but they are actually cultural design projects.

To this end, a central question has come up: is the aesthetic aspect of our work the most trivial or, unstated as it is, could it be the most powerful? Could conscious design be the thing that sets landscape architecture apart from science or engineering? In our increasingly virtual world, can an explicit focus on an aesthetic agenda and material design have the potential to cut through a disengaged experience of the physical, to shape people's perception?

Mikyoung Kim: Maybe we are talking more about process and creativity rather than aesthetics. At the summit, we saw analytical, scientifically based diagrams that are so beautiful to us because they are methodical, conscious, and backed by scientific data, whereas the creative process or creative process of thinking relies on intuition...which comes from our subconscious. Analytics on its own is not enough to create great design. You need both the conscious and the subconscious; we need to use both hemispheres of our brain to create great work.

If we veer too wildly to one side, the data-based design, we fall into the danger of creating designs that may lack the uniqueness that we as individual, creative beings bring to the profession. And if we veer too wildly to the other side, we wallow into the subconscious and the intuitive and create work where we ignore all the pressing matters. Landscape architecture is both the art and the science of design and we need the declaration to reflect that.

Chris Reed: When taking on ecological, environmental, social, political, and economic agendas, we need to lead with design first. As a design profession, we have an obligation to push a strong and central cultural agenda. And we need to address important matters through design—the qualitative and the experiential—and do it in contemporary ways. At the end of the day, in order to get at these issues, we need to do it by moving people's hearts, triggering their imaginations, and stimulating their minds.

Maria Goula: I like the idea of reordering because it is in the core of design and has special systemic and processual implications. We do not simply need new aesthetics, but maybe multiple aesthetics.

Ken Smith: You cannot consider the question of aesthetics without addressing issues of content, meaning, and intentionality. Aesthetics are the qualitative aspects of design that allow a designed place and its spaces, program, and forms to speak and provide meaning to human users. It may delight, confound, enlighten, sadden, soothe, or exhilarate. Aesthetics are the emotional core of what we do as landscape architects. Clearly, we live in an increasingly data-driven culture with an emphasis on quantitative metrics, prescriptive models, minimum standards, and best practices for almost all aspects of landscape performance, often to the exclusion of the qualitative concerns. Through digital means, it is easy to appropriate precedents, adopt models, and apply parametric algorithms to generate workable design solutions. Rhetorically speaking, with the prospect of artificial intelligence, one might ask, why do we even need designers and aesthetics? Can we simply model the quantitative metrics to solve the problems? And is fabrication simply the application of design standards and specifications?

Intentionality is what gives our work meaning and direction. Our aesthetics, philosophical position, interpretation of the program, working methodology, along with questions of structure, proportion, materiality, process, and fabrication, are critical to our craft and fundamental to our art. Aesthetics lie at the heart of our professional mandate. It is how and why we create meaningful places that matter to people and matter over time.

Claude Cormier: The issues are extremely complex, but this is where an intuitive element comes into play: you align

everything and the whole journey of the project is trying to keep all those things streamlined. Because in the end, when you do the act of design, you commit. And this commitment is expressed in form; and form takes time to get realized. You must have a very clear intention, higher than all of those issues, because if you just worry about the issues, in a sense, you are going to sink with them. Design helps you start floating.

Chris Reed: The discipline has made amazing strides in recovering a central and more complex role for what we can do in cities and society, but there is still resistance to engaging at the level of design experimentation, aesthetic values, and aesthetic qualities. Right now, architects still hold more cultural capital than we do. I challenge us to take on these issues, have these debates, put them front and center, be more experimental, be more out front, and hopefully we can raise our own cultural capital.

Ken Smith: A lot of times we are trying to solve contemporary environmental issues and social issues with old aesthetics. The art world does not speak with a single voice—it has multiple voices and is experimental, and it seems to me that we should be like the art world, in terms of having a panoply of aesthetics. We need to be developing contemporary aesthetic solutions to contemporary problems.

Mikyoung Kim: Most of us have a global practice and it is our obligation to not just bring a Western aesthetic when we are working in eastern Asia or in the Middle East; we have to be more open and collaborative.

Maria Goula: Good practice has always been collaborative. But there was an old paradigm to venerate or exaggerate the genius and the talent of the artistic figure leading the process. We have moved into very different horizontal processes with negotiation and constant dialogue and a lot of frustration, of course, and energy. But no one imagines working in a perfect world of having aesthetic creative production on her or his own.

Adam Greenspan: Form, design, and aesthetics need to get into the declaration because it is an important part of what we do in the world.

Chapter 36

ECOLOGY PANEL

PANELISTS

José Almiñana, Principal, Andropogon Associates

Julie Bargmann, Founding Principal, D.I.R.T. studio; Chair and Associate Professor, Department of Landscape Architecture, University of Virginia

Brett Milligan, Assistant Professor of Landscape Architecture, Department of Human Ecology, University of California, Davis; Director, Metamorphic Landscapes

Ellen Neises, Founder, RANGE; Adjunct Associate Professor, PennDesign, University of Pennsylvania

Antje Stokman, Professor and Director, Institute of Landscape Planning and Ecology, University of Stuttgart

MODERATOR

Kristina Hill, PhD, Associate Professor of Landscape Architecture and Environmental Planning and Urban Design, College of Environmental Design, University of California, Berkeley

Ecology is the interrelationship of organisms and their environment. The Ecology panel was asked to discuss its status within the landscape architecture discipline in an age of climate change. The following is a condensed and edited version of the panelists' remarks and is not intended to be a verbatim or complete representation.

Kristina Hill: My everyday work is about climate change and thinking about adaptation and what we learn about science and predictive models. I hope we get a stronger sense of urgency because we might look back on this point in time and feel that there were things that we knew that we did not take as seriously as we should have when we heard them.

Summit participants were polled on the level of urgency that exists now that the environments we have altered are forming a threat to us. More than 60 percent of participants said we still have time before that new crisis begins and we can avert it by doing what we have done in the past. And 34 percent said we are in the early stages of this new crisis and must prepare our profession to act in new ways that are appropriate to a new era. A mere 5 percent thought it is already too late for us to avert major catastrophes.

Personally, I would disagree that we have time and we can do more of what we have been doing up to this point. At the very least we are in the early stages of this new crisis and must prepare our profession to act in new ways that are appropriate to a new era.

Ellen Neises: Culture shifting practices are at the heart of everything. Ecology is never in opposition to aesthetics or poetics until our gorgeous landscapes start creating a false sense of well-being and mask the anxiety that we feel and drain the rightful content out of landscape, inhibiting public consciousness and creating a kind of sustainability propaganda. Optimism sells, but design has to deliver less pacification and much more provocation.

José Almiñana: Every time we undertake a commission, we are given the responsibility and opportunity to be transformative. I would like to pose a series of considerations for that time. Does this project increase the complexity of the habitats it creates? Is it entropic or negentropic? Does it align resources and processes that sustain life? Is it biophilic? Does it promote the integrity of the community where it is located? Is it just? Does it promote the life associations that are native to the ecoregion where it belongs? Is it designed to last? And most importantly, is it able to adapt to change?

For all of this to happen, we must embed scientific rigor in our profession. We have no clue if the transformations we are making are effective. So, measure the performance to test your hypothesis, design your experiment, and find out that if what you are doing matters.

Brett Milligan: I would challenge our profession to move past nature because it is incredibly subjective. Ecology forces us to think about agency, including the collective agency. Ecology brings us to a relational ontology of landscapes. Ecology is inherently political. We need a more general notion of ecology. It is not just biotic material—it is sediment, bricks, copper wire, us, a relational way of looking at landscapes. When we talk about ecology, we tend to see the landscape as some sort of backdrop that is passively taking what we do to it. We do things to landscape and that collective agency responds to what we do in ways we do not like or ways that complicate our efforts.

Julie Bargmann: We need a declaration of *urgency* more than a declaration of concern. Our practices should be more than interdisciplinary; they should be interdependent, just like the issues.

Kristina Hill: We need to be aware of the impacts of territorial conflicts, climate change, and resource conflicts on human populations. The United Nations estimates there are more than 60 million refugees in the world right now. How is the movement of refugees going to affect us, our politics, our ecologies, our ecosystems in the biophysical world, and our own abilities to summon compassion?

Antje Stokman: We are having debates in Germany on how to accommodate the influx of refugees. Citizens have tied themselves to trees, defending what they thought of as their right to the environment against new refugee shelters being built in my neighborhood—which shows the entanglement of human issues of ecology related to biotic issues of ecology. We need new ways of organizing community—how people live together—and of engaging ourselves in those transformative processes. We must define culturally specific ways where

people exchange knowledge, learn from each other, exchange globally, and find answers that are specific to the cultural context of where they live. We need to develop new ways of negotiating, experiencing, and storytelling that guide landscape architects to overcome the growing social conflicts in our cities.

Kristina Hill: In an urban region, we have cyborg biodiversity—species altered by human action, such as bees that we have genetically altered. We live in an altered world of biodiversity, not a precious world of only species that exist outside of human influence. We have influenced every aspect of biodiversity and must begin to think about it in a new way, not the old native/nonnative way, but understanding the influence of people on biodiversity.

José Almiñana: If we understand that we are in the business of sustaining life on the planet—that has to find a place in the economic model that governs an econocene. The concept of ecosystem services was created in the nineties and is the underpinning of the SITES rating system. It is the notion that every site can provide the value that sustains life on the planet.

Antje Stokman: The Global South is where these challenges are occurring. We need to redefine our thinking about how to cooperate. We set up a program funded by the German and Egyptian governments to work around the Mediterranean Sea.

The students spend the second year in Egypt where they can work deeply in very challenging, complex environments while doing their master's theses. In Egypt, we can relate directly with the local communities and work in the field. Those of us from the Global North learn much more there than we learn here. And we have had enormous impact.

Kristina Hill: We must figure out how to build capacity for countries that are going to have billions of people and do not need us to save them. They need us in the developed countries to do the best we can *not* to make things worse for the rest of the world. The separation of ecology from other issues today is not something that we see as intellectually valuable. We use that word *ecology* as a bit of an anchor and then ask, "What relationships, what species are we talking about?" And we need to squarely include the human species in that conversation, with all the ethical and philosophical issues that brings up.

Chapter 37

SOCIETY PANEL

PANELISTS

Diane Jones Allen, Principal, DesignJones LLC

Alison B. Hirsch, PhD, Assistant Professor, Landscape Architecture and Urbanism, School of Architecture, University of Southern California; Cofounder, foreground design agency

Jeffrey Hou, PhD, Professor and Chair of Landscape Architecture, Department of Landscape Architecture, University of Washington

Anne Whiston Spirn, Professor, Landscape Architecture and Planning, School of Architecture and Planning, Massachusetts Institute of Technology; Director, West Philadelphia Landscape Project

MODERATOR

Deb Guenther, Partner, Mithun

Because landscape architecture reflects culture and has the power to change communities and shape people's lives, landscape architects must consider societal issues. Diversity, alternative modes of practice, and cultural nuances were among the topics that the Society panel addressed. The following is a condensed and edited version of the panelists' remarks and is not intended to be a verbatim or complete representation.

Deb Guenther: We talked about the need to become policy builders, to challenge the diminishment of public space and our rapid change in culture due to globalization and mass migration. We focused on three key points: acknowledge and inspire diversity, be proactive by making design and equity opportunities happen, and radicalize and be willing to get into the "mud of politics." Diversity is an important way that we do our work and those voices are important.

Diane Jones Allen: To address the social, cultural, aesthetic, and ecological challenges that we face, we need diverse minds, a diverse population, and a diverse set of landscape architects. But first and foremost, we must recognize the diversity that we already have within the profession. Landscape architecture is a global profession with people of very different colors, ages, and races; we should be acknowledging that they are already there. For example, there was an African-American landscape architect president of ASLA, an African-American landscape architect Presidential Medal winner, and an African-American landscape architect working in the neighborhood where Freddie Gray died who was using landscape architecture to heal that community. If we really want diversity, we need to recognize it, praise it, and lift it up.

Jeffrey Hou: All the grand challenges that we face in the next 50 years—sea-level rise, acute flood events, storm surges, migration, urbanization, and so on—are happening in different contexts and impacting different people. Technical solutions are important, but we need to diversify our thinking to understand the political and cultural nuances of how solutions apply to different contexts. If not, this profession will become a form of colonialism, a colonial practice that imposes order and structures and norms on others, especially vulnerable populations, and a kind of colonization that assumes there are universal solutions and tries to impose them on others.

Anne Whiston Spirn: In 1970, this profession did not look anything like it does today. There were very few women and very few students with liberal arts degrees. Look at our profession today and the vast numbers of women. The influx of folks with liberal arts backgrounds also helped transform the profession from 1970 to what it is today and is an example of how diversifying our ranks can enrich the profession.

Deb Guenther: Our second issue is about being proactive and about making design and equity opportunities happen that do not always fit within our traditional practice modes. Traditional work processes can be supported by alternative modes of practice.

Anne Whiston Spirn: Around 1984, I decided to devote a lot of my work to looking at how to bring together restoring the natural environment and the urban natural environment with rebuilding inner-city communities. And how to do so in ways that beautified the communities, as well as made them healthier, safer, and more just. I went into the West Philadelphia neighborhood and it became very clear to me that it took a long-term commitment to build trust in the community. I originally planned to stay only for the four years of the grant I had, but I realized that I had built many relationships and that made so many other things possible for me to investigate that would not have been possible before. This was a community that did not know they needed a landscape architect.

I view my own role as an academic practitioner to open up territory, to demonstrate the efficacy and importance of landscape architecture thinking and what landscape architects have to contribute to places where people do not think of landscape architects as having a role.

Jeffrey Hou: We should be recognizing multiple voices and supporting and advocating for those who are not at the table. Unfortunately, we did not see that in the 1966 declaration. The profession missed the opportunity 50 years ago; let us not miss it this time around.

Alison Hirsch: A number of practices today are really meaningful exceptions to the continued tendency of the profession to avoid issues of inclusiveness. Those practices do not necessarily sacrifice the qualitative project. And they do not leave these efforts to those few community-oriented design practices, but integrating the social component becomes integral to the way we practice today.

A number of people advocated for cultural landscape preservation and safeguarding cultural heritage, and one way to interpret and expand on that is to recognize that place and cultural identity are no longer so intimately bound but are increasingly contested in a world of mass migrations of people. It challenges and complexifies the ecumene. Part of recognizing the localization of the global is bringing forth new hybrid acts of occupation and expression in public space through meaningful localized intervention. And that might be one means of mitigating the idea of the potential of cultural colonialism.

Deb Guenther: Our third point is to embrace the idea of radicalizing our profession and getting more involved in taking a position. For future generations, it will be a requirement to be interdependent; how do we help make happen in academia?

Diane Jones Allen: Get students into the community so communities see what landscape architects can do and students understand the diversity of communities and issues.

Jeffrey Hou: But if we bring students out to a community and our methods and pedagogy do not change, we are back to cultural colonialism.

Anne Whiston Spirn: The last 30-plus years of action research have taught me the importance of doing a physical program and a social program, together, so every social program should have a physical manifestation that can serve as a symbol and inspire and attract students.

Over time I moved from advocacy to capacity building. I still advocate, but part of it is empowering and leaving the neighborhood not only with a better plan or a better design than existed before but also with more knowledge and capacity building to move on and continue. This is a way to transform practice—so it becomes a colearning process and it transforms your own practice.

Deb Guenther: How do we attract individuals from communities of color or less privileged upbringing into the profession?

Anne Whiston Spirn: We go into schools. I worked on an urban watershed curriculum with a middle school in West Philadelphia for six years and those kids learned what landscape architecture is.

Diane Jones Allen: We need to get more of us in school, make more of us professors, have more of us owning practices. But we are out there; our organizations must expose those of us that are out there working so that we can increase our capacity to deal with these problems, these diversity issues.

Chapter 38

INNOVATION PANEL

PANELISTS

Andrea Hansen, Principal and Creative Director, Fluxscape; Head of Product, stae.co

Liat Margolis, Assistant Professor of Landscape Architecture and Director, Green Roof Innovation Testing Laboratory, University of Toronto

Adrian McGregor, Managing Director, McGregor Coxall

Marcel Wilson, Founder and Design Director, Bionic

MODERATOR

Karen M'Closkey, Associate Professor, University of Pennsylvania School of Design; Cofounder, PEG office of landscape + architecture

Innovation is a buzzword that seems to mean different things to different professions. This panel discussed what they see as innovation in landscape architecture and how it can be fostered. The following is a condensed and edited version of the panelists' remarks and is not intended to be a verbatim or complete representation.

Karen M'Closkey: The word *innovation* was used in half of the declarations, often generally. And it was assumed that as the world changes, design must change with it, or more importantly, that design can provide visions for how it can change. What constitutes innovation cannot be determined in general terms; it depends on context. Technology itself does not create design innovation, but it can foster techniques that open up design possibilities, such as computer simulations that can offer a relational and time-dependent means for working with a dynamic medium.

Data visualization itself—who produces it, who has access to it, and how it is spun—is one of the most important tools for telling a story. The creative combination and synthesis of diverse scales and sources of information enable us to perceive and articulate broader issues of the material, social, and cultural environment.

Adrian McGregor: The twenty-first century is going to see a quantum shift in the way that landscape systems are valued. As the global population heads toward nine billion this century, competition for our remaining virgin environmental resources is going to trigger profound changes to economic theory and practice. We are going to see a new economy and under that new economy, the trading of carbon itself, because carbon is the kernel of climate change. I think carbon is eventually going to manifest itself as a global market and it will be underpinned by the monetary quantification of ecological

and landscape values. Around 60 carbon-trading programs are already running in countries and individual locations. And eventually they will come together and harness the internet of things: big data, real-time management.

The outcome is that it will trigger a flow into smarter design practices that optimize life-cycle efficiency across energy, water, and carbon systems. Design and creativity will still play a crucial role in these data-enabled project workflows. We still need design thinking to drive the disruption and the innovation, but research and education practice must fall into line. If landscape architecture wants to play a significant role in this emerging space and emerging economy, we need to change, to develop tools that quantify embodied energy and operational carbon, and we need to find the way to display those in our public spaces and environments.

Andrea Hansen: The complex global problems necessitate an innovation in design practice—an expansion of our profession's focus beyond just site design—to reflect our unique capacity for design thinking. Landscape architecture itself has to innovate and expand in order to better prepare our next generation to spread the beneficial principles of landscape architecture. We can encourage students to pursue careers in fields outside traditional landscape architecture— research, policy, government, journalism, visualization, or a technology start-up. We must do a better job of fostering individual agency, teaching young landscape architects to be their own advocates, to craft their own agendas, and to find their own voices, thereby training them not to be just good employees, but also proactive entrepreneurs.

Marcel Wilson: The industry of landscape architecture is largely consumed with patching, fixing, and covering stuff up and making the best of it—designing to the minimums. And landscape architects as a result tend to be reactionary, led by constraints, and risk averse.

In the broadest view of the industry, there is very little will or the skill to incentivize innovation—a relatively new ambition for the profession. If you look at other industries and fields, significant shifts in innovation are intentionally stimulated. Agriculture, medicine, aerospace, computer science—funding goes in, investments are made, teams are formed, resources are leveraged, and the conditions are created for great leaps to be made, and eventually, the results are applied to government applications, trials are conducted, and then they are adopted by industry and the public sector. Right now, government and industry do not fund landscape architects to invent solutions to grand challenges facing humanity. Technology-based landscape innovations at all scales and applications just have to accelerate.

Liat Margolis: Where are our think tanks? Where are our technological incubators? What shall we define as disruptive technologies and practices in landscape architecture? We have mobile technologies that have given rise to a huge social transformation in social media and disruptive economic structures such as Uber. Think about our contribution as an academic and professional field to experimentation, prototyping, and research and development; to interdisciplinary knowledge transfer and education; and to linking academic research within the public and private sectors.

Real-time, continuous, time series data are available—whether via sensors or field technology. How do we think about dealing with flood management or fire prevention in extreme climate events and how do you integrate these kinds of actuation into a real-time response of landscape?

The other dimension is long-term thinking: how do you archive performance over time and analyze it and understand in the long term how well we are performing? And how do you provide that feedback to designers, operators, legislators, industry standards, and user occupants? This provides both a quantitative and qualitative comparison and evaluation of the design intent—what is called preoccupancy and post-occupancy—or actual performance. What would it mean if the workflow did not end when the project was constructed, but rather this kind of feedback loop would be for the lifetime of the project? It changes the role of the designer within the entirety of the lifetime of the project.

Adrian McGregor: We must get to a point where the data around the project—the preemptive predictive modeling done on the design side—actually give you a gateway for approval. If the project is too consumptive, it does not get approval. Carbon trading is where the market will go because if you build a project in the city, you cannot offset all the carbon. There are only so many plants, green roofs, green walls, and trees you can put on your site. The offset might come from the Amazonian rainforest or the Colombian coffee plantations. All of a sudden, they have a quantifiable value and are in the value chain. Not only do they have their agricultural value, scenic value, or other intrinsic values, but they also have economic value because that can be traded and offset against what a project needs to be delivered. That is the loop.

Andrea Hansen: Innovation is perhaps a highly positivist term that still does not recognize the benefits of *failure*. So long as we are failing forward and learning from the mistakes that we make and doing that in a safe environment, we should not be afraid of the term *failure*. We should think about the ways that an incubator or accelerator or venture capital model might free us from the constraints of a client/consultant model.

Chapter 39

ACADEMIC PRACTICE PANEL

PANELISTS

Anita Berrizbeitia, Professor and Chair, Department of Landscape Architecture, Harvard Graduate School of Design

Jacky Bowring, Professor of Landscape Architecture, School of Landscape Architecture, Lincoln University

Julia Czerniak, Associate Dean and Professor of Architecture, School of Architecture, Syracuse University

Susan Herrington, Professor, School of Architecture and Landscape Architecture, University of British Columbia

Anuradha Mathur, Architect and Landscape Architect Professor, University of Pennsylvania School of Design

MODERATOR

M. Elen Deming, DDes, Professor, Department of Landscape Architecture, University of Illinois at Urbana-Champaign

The Academic Practice panel was charged with discussing issues from the declarations and panels relating to higher education, including teaching and research. The following is a condensed and edited version of the panelists' remarks and is not intended to be a verbatim or complete representation.

M. Elen Deming: Academic practice permeates private and public practice as well as advocacy practices. Many of us on the panel are, have been, or will be administrators, landscape professionals, community activists, competition winners, and founders of our own design offices. Nearly all professional programs hold faculty accountable in three domains: teaching, scholarship, and service. Yet in a much larger sense, academic practice is simultaneously normative, constitutive, and transformative of its discipline—any discipline. And as academic formats and cultures vary, this larger purpose gets mapped onto our teaching, scholarship, and service in a hundred interesting ways, with variable impacts. The design of curricula shapes learning outcomes, which impact future professionals. Research agendas also shape new knowledge, which then affects core and emergent societal needs. And the transfer of knowledge through outreach or engagement impacts social and intellectual capacity for stakeholders as well as professional communities.

Jacky Bowring: I am interested in definitions and the power and danger in the language that we use. We need to take care with what is bound up in our words. A first question about words is: who are we? Context, identity, and belonging are core aspects of landscape architecture. The declaration must be precise in terms of *who* is declaring because what you see depends on where you stand. A second question is about the biases of the declarations presented at the summit, which show an emphasis on the macro scale and science. Bound up in the question of scale is another question of context. A singular focus on urbanism has been repeatedly invoked in the declarations, and a statement that future priorities for landscape architects will all center on the city is chilling. Coming from a country (New Zealand) where design priorities lie across the entire range—from wild places to the urban context—this again raises a question about where the declaration finds its home, its heart. Beyond the muscularity of insistent urbanism, where is the place for humility, fragile design, and the setting aside of egos?

Susan Herrington: We have heard a lot about science and how it is supposed to redeem the profession. We must remember that scientific fact alone does not change human action or behavior. Culture changes culture. In a landscape architecture program, we spend a lot of time in design studio, on visualization, communication, and the technical parts; we are not trained as scientists. However, we can collaborate with scientists and bring those skills to the collaborative process.

Julia Czerniak: Academic practitioners are crucial players that reflect on the discipline and the profession's role in addressing provocations. The highest function for academic practitioners is to challenge both conventional thinking and conventional things. Through designing, we can test new materials, fabrications, and forms in safe-to-fail environments, expanding on our technological capabilities to support designing for dynamic landscapes. Through writing, with the privilege of a bit more time, we can advance design strategies that bridge thinking and making, as well as frameworks to inspire action.

Anita Berrizbeitia: Academia has been the great incubator of ideas in the field and that is a core value that we take seriously. We practice from a critical stance in which methods, assumptions, identities, and the very limits of knowledge are questioned, actively reconnecting back to the conditions in the world. Collaboration is increasing with new forms of partnerships between industry and academia, and nonprofit organizations and such. Collaboration and interdisciplinary work are key; at the same time, we need to understand that interdisciplinarity is something that works to expand what we do.

Anuradha Mathur: The declarations tend to position themselves at the end of history. They expressed the belief that we now have the right view of the land, of environment, of ecology, of culture, and therefore, are in the right ballpark of solutions. While many of these declarations also articulate the need for more knowledge, data, science, inclusion, scale, design, wildness, and politics, they do not question ontological underpinnings—the things we take for granted as existing as real, as natural, as being. It is not enough for designers to be

grounded in things that are known to exist and exist as known. They need to posit new things and test their possibilities, not just for a future, but for reconstructing a past, as well as re-articulating our present.

This is design that is beyond the facts and critiques of empirical and critical research. It assumes that people do not see things differently but that they see different things. In a milieu where professional practice and academic practice are becoming increasingly complicit in maintaining design as an applied field, we need to promote design as a field of inquiry.

M. Elen Deming: It seems that academia is the first line of defense against the end of the world as we know it. According to the declarations, innovation in landscape architecture seems to rely largely on new methods, new tools, drone technology, new techniques. And if design methods, tools, and techniques are our science, then perhaps we need to double down on that.

Anuradha Mathur: Innovation is always coupled with new technology, but innovation does not always have to be connected to technology. It is also relooking at ways of seeing and being critical.

M. Elen Deming: There is a value in pragmatic research, and there is a variety of ways to do that. One way is to partner with your local university at almost any level. These synergies can stir the pot between the kind of pragmatic, practical performance problems that you face in your offices and help to refresh and make the academic agenda as relevant as it can be to the discipline.

Anita Berrizbeitia: Do we need to change the fundamental educational construct to reduce financial barriers to broad access to the profession? There are a lot of structural challenges. It is not just about what education costs but a larger question of what value society itself places on landscape architecture. The great challenge that we face is that everybody loves beautiful spaces and everybody loves cities, dynamic cities, filled with fantastic public spaces, and for some reason they are not willing to pay for them.

M. Elen Deming: In summary, we see the role of academic practice to emphasize technique and speculative ways of seeing and making, to integrate, advance speculative research, engage the potentials of criticism, but also, very importantly, champion the socio-environmental justice in an epoch of perceived and constructed scarcity. And finally, to provoke. And that means to lead the profession by taking political and intellectual risks.

Chapter 40

PRIVATE PRACTICE PANEL

PANELISTS

Thomas Balsley, Founder and Principal Designer, Thomas Balsley Associates

Keith Bowers, Founder and President, Biohabitats; Founder and Partner, Ecological Restoration and Management, Inc.

Joe Brown, Planning and Design Consulting Adviser, AECOM

Kathryn Gustafson, Founding Principal, GGN

Mark W. Johnson, Cofounder, Civitas

MODERATOR

Laura Solano, Principal, Michael Van Valkenburgh Associates

The Private Practice panel discussed ideas from the declarations and other panels as they relate to the work of design firms. The following is a condensed and edited version of the panelists' remarks and is not intended to be a verbatim or complete representation.

Laura Solano: Private practitioners are at once synthesizers and dreamers. We are willing to have our ambitions, skills, talents, and vulnerabilities displayed in public without being able to be there to explain ourselves. We have tremendous responsibilities. In our projects, we aim to move people, make them happy, make them feel safe, and add to their daily lives. With our clients, we are charged with fulfilling their ambitions and doing it on schedule and on budget. We go out into communities to convince public agencies and citizens that our work is in their interest. We train, motivate, and invest in the next generation of practitioners, and we continue to educate ourselves, and hopefully, run profitable businesses. And we aim to do all this without harming the earth.

Mark Johnson: Think about what doctors do. Their goal is to sustain life. They all know what they are trying to do—whether they are a researcher, a clinical doctor, or wherever they are within that very diverse field. But landscape architects do not share that sense of common purpose. Our academy could be that way; we could become like doctors, aligned but richly separated by our specialties. Doctors think about the Hippocratic Oath, which is now 2,400 years old. What if our oath was to sustain life in all its forms? Something as simple as a healthy planet for healthy people? Then, for example, if you were doing a grading plan, you could ask: does this improve the health of people or the earth? Am I doing good or doing harm?

If we had that simple ethical basis and common purpose, we could teach people coming into our field to look at the consequences of their actions. We need to teach and remember that design has to be accountable to everything in the world.

Laura Solano: Large multidisciplinary firms are joining forces and they seek to buy out smaller firms all over the world, yet many firms have fewer than 20 people. How do small firms make an impact? How do they be persuasive and make a difference?

Joe Brown: Large firms, such as AECOM, can have silos of civil engineers, landscape architects, architects, scientists, and others. Isolation can happen regardless of size. You need to reach out and communicate with everything going on in your community, to exercise curiosity and be expert-driven: smart people seek out smart people.

Thomas Balsley: The key is collaboration. In order to feel that you are making a difference and an impact, some of us—especially the smaller firms—believe that we must be in the big game, dealing with big problems and issues. Our studio is relatively small, but we have done very large work by collaborating with another firm. The math we use at our job interviews is that one plus one equals three.

Laura Solano: How do landscape architects bring the full force of our talents and environmental credentials to the table, and capture a leadership role within a multidisciplinary team, even when we are not leading those teams?

Kathryn Gustafson: When you walk into a team, you have to know exactly who all your experts are and demand from them what you need to make the project work. You must be proactive and have confidence in the multiple talents you bring to the table—confidence that you are the glue, the thing that pulls everything together. You need to take that leadership role. You also must do your homework. Set your goals and know that you have to be the smart person in the room so everybody listens to you, and that way, the landscape will lead.

Mark Johnson: You do not become a leader by being a generalist. We must know something other professions do not. We must have our feet on the ground with our knowledge base that others do not fully share, just like we do not fully know and share theirs. On a team, you cannot be afraid to ask questions. You need to ask questions to get everyone to question themselves. As you do this, you will be challenged, and you will get better at it. You will ask more clever questions the next time and learn more about how they think as they rebut what you are doing. Pretty soon, you will find that you are the person framing the situation for everyone else, so that

you are the integrator—not of the activities, but of the issues and ideas. Once you do that, you are the leader.

Joe Brown: It is not so much *leading* as *listening.* Listening by leading, leading by listening.

Laura Solano: Let us move on to resiliency and projects.

Kathryn Gustafson: For a landscape to actually last, it does not just last ecologically—it has to last in the hearts and minds of people. Beauty is different from aesthetics. Aesthetics is a style, but beauty has a soul and connects with you emotionally; it elevates into a higher level of arousing your senses and your thoughts and spirits.

Keith Bowers: We need to look at a way where a vibrant and robust economy is totally dependent on a healthy, just, equitable and fair society. And to achieve that, we need healthy, intact, and interconnected ecosystems. Without that, resiliency is not really going to work. That is also true in the way we operate our companies, our practices, our firms, and everything from how we treat people, to the types of work we go after, to the types of relationships we build. It all needs to be nested or interconnected on both physical and temporal scales to be resilient.

Resiliency starts with wildness, nature. If we do not have wild nature as a juxtaposition or a balance to cities and urban areas, we do not have resiliency in the long run and our cities will fail. We need to be leaders in our urban work and the rewilding, reconnecting, and restoring work that is desperately needed in order to have those intact ecosystems that keep us all healthy and alive.

A second tenet is ecological democracy. We in the United States sometimes forget that we are consuming resources at an unprecedented rate. And as people in the developing world move into the city and their living standards go up, there is going to be an even greater impact on our resources. We must find ways to work in those realms.

Laura Solano: What is the future responsibility of practice? What will change?

Joe Brown: What will change is change. Change will get faster worldwide.

Mark Johnson: We must be responsive to, and responsible to, the world being driven apart more quickly by cultural and political narrative than perhaps even climate change and the other big things that we face. And if we do not learn how to enter our voice into that narrative, whether in a political way or a cultural way or both, we are going to lose.

Kathryn Gustafson: We have to find a way that educates the public to see what we do, and to see that it has taken a human hand to make it work better. It is not just nature taking care of itself; we are taking care of it.

Thomas Balsley: We were all attracted to this profession because of that first word: *landscape*. But we can work at these challenges better if we pay as much attention to the policies, the people, the art, the social, the economic—all the issues—and wrap our arms around them with the same kind of embrace as we would a tree.

Joe Brown: We also need to get into the developing economies of the world because that is where the danger is; that is where we will fail the earth as our client.

Chapter 41

PUBLIC PRACTICE PANEL

PANELISTS

Nette Compton, Senior Director, Park Central and City Park Development, Trust for Public Land

Mark A. Focht, Deputy Commissioner/Chief Operating Officer, New York City Department of Parks and Recreation

Christian Gabriel, National Design Director—Landscape Architecture, US General Services Administration

Ed Garza, Founder and Chief Executive Officer, Zane Garway

Deborah Marton, Executive Director, New York Restoration Project

MODERATOR

Mia Lehrer, Founding Principal, Mia Lehrer + Associates

The Public Practice panel discussed ideas from the declarations and panels relating to the government and nonprofit sectors dedicated to serving the public good, stressing the opportunities for landscape architects.

The following is a condensed and edited version of the panelists' remarks and is not intended to be a verbatim or complete representation.

Mia Lehrer: These panelists work at the intersection of people, place, and policy, making our communities healthier and more livable. If we look at landscape architecture as having three realms—projects, programs, and policies—it is the built projects that get the most recognition in the field. Those working in the public realm, whether developing programs, changing policies, or serving in public office, may feel disconnected from those who call themselves landscape architects. And yet our public practitioners are at the forefront of emerging research, urban transformation, groundbreaking policies, and communication with the public.

In addition, they often serve as curators, collaborating with practitioners to deliver projects. Our panelists here are infiltrating organizations that many of us have as clients and are elevating the discussion around landscape architecture. They are basically defending us, expanding our roles, and raising the bar of the work that is being produced.

Deborah Marton: The broader culture, as many have said, does not really understand what it is that we do. By taking our perspective and becoming the client, the developer, the nonprofit leader, the mayor, and by leading, we can change the broader values to create a kind of conceptual and ideological framework for the work that we can deliver.

Nette Compton: The power of our profession is not just to advocate for large, global-scale issues like climate change but to create places of experience that stick with people for their lifetimes because that is how you create a population that cares about these bigger picture issues. Without the foundation of person-to-person experience, individual experience, and transformative experience, you are not going to translate to an adult who votes on these issues or wants to fund these issues.

And when we talk about the importance of expanding our profession, if we focus more on convincing the public in the day-to-day places and the quality of life that they experience through our work, we will drive enough people to this profession just as an offshoot of that effort.

Christian Gabriel: To the degree that we actually move into roles of greater authority and really reclaim that word *landscape*, we can move beyond the boundaries of park design and start thinking about the more aggregate components of what planning is in design and construction, and how we can effect change in actionable, measurable, and proactive ways.

We can effect change in any number of ways—through policies, measurable project-level results, landscape analytics, and looking at how data inform processes so knowledge-based design and decision-making really start factoring into the discussions.

Policy is a mechanism and a vehicle to deliver action as a proxy across projects where you cannot always be in attendance for every little decision and meeting. Once these things get recaptured and reconceptualized into the mantra of ecosystem services, that immediately connects us to our allied professionals, but also allows us to somehow be differentiated and gives us room to bring delight to projects. That is where our greatest future and capacity are.

Mark Focht: I think landscape architects are inherently wired to work in the public sector. Everything in our education, training, and practice is about collaboration and building consensus. My challenge to everyone is to work in the public sector, really influence policy, practice, and the finished product, and use what we are inherently wired and trying to do to help benefit the environment and communities.

Nette Compton: Nonprofits have an opportunity to fit in between the needs of a community and the capacity of government structure. And that interstitial space varies at different scales and in different cities. Our skills as landscape architects, collaborators, listeners, conveners, and ultimately synthesizers of solutions enable us to identify what the opportunities are and to bring together a diverse suite of partners from funders, the community, and the government to bring projects to fruition.

Students should think about judging the *success* of their work on the *impact* of their work. That is a different value system than is often found both in schools and in our practice.

Ed Garza: We need more landscape architects to be in the position of elected office, whether it is the school board, city council, mayor, or all the way up through the federal level. If everyone in here ran for office, you would begin to see that side of the table advocate for better projects. But it has to be supported on the grassroots side. People need to be educated as well as the leadership.

Mark Focht: It is uncomfortable to most people, but the way we are going to effect change is to place ourselves in the political structure. Start small—your zoning commission, your shade tree commission—but start somewhere and become a landscape architect that is in the public realm representing the broader good. Resources today are being driven at the local level. Most innovation is happening at the local level.

Mia Lehrer: In addition, become members of boards of nonprofits like the Trust for Public Land, TreePeople, or the Nature Conservancy.

Deborah Marton: Something that I think young professionals should find very attractive is that, if you work with any of the institutions that we are leading, you can do a lot at a very early stage. Young landscape architects who work with us are meeting with the community, doing schematics, doing the full drawing set, overseeing construction; five years in, they are building their own landscapes.

Ed Garza: We have 20,000 incorporated cities in the United States and I would dare to say at least 60–70 percent do not have a landscape architect that lives in their communities. If we are going to become the change agent and the leader, then we have to grow the young professionals. We must reach out to the schools to help them understand why this is important, because there are too many communities that do not have voices for these important policy issues.

Nette Compton: Within the nonprofit and government realms, we need to demand excellence in our most underserved communities. These are the communities where the intervention will be most transformative, and we need to be investing in those spaces and demand excellence in those spaces, more than any other work that we are doing.

Chapter 42

CAPACITY ORGANIZATIONS PANEL

PANELISTS

Barbara Deutsch, Chief Executive Officer, Landscape Architecture Foundation (LAF)

Kathryn Moore, Professor of Landscape Architecture, School of Architecture and Design, Birmingham City University; President, International Federation of Landscape Architects (IFLA)

Raquel Peñalosa, Raquel Peñalosa Architectes du Paysage; Delegate and President, IFLA Americas

John Peterson, Founder, Public Architecture; Curator, Loeb Fellowship, Harvard Graduate School of Design

Patrick L. Phillips, Global Chief Executive Officer, Urban Land Institute (ULI)

Nancy C. Somerville, Executive Vice President and Chief Executive Officer, American Society of Landscape Architects (ASLA)

MODERATOR

Stephanie Rolley, Professor and Department Head, Landscape Architecture and Regional and Community Planning, Kansas State University

The Capacity Organizations panel was asked to explore how professional membership organizations and nonprofits can help shape the future of the landscape architecture profession as related to the issues highlighted in the declarations and panels. The following is a condensed and edited version of the panelists' remarks and is not intended to be a verbatim or complete representation.

Stephanie Rolley: We are talking about the future of landscape architecture through the eyes of professional organizations and who represents us and how we have a larger voice. The International Federation of Landscape Architects (IFLA) represents over 25,000 landscape architects in over 70 countries. The American Society of Landscape Architects (ASLA) is a member of IFLA and represents over 15,000 members. The Canadian Society of Landscape Architects (CSLA) represents over 2,000 landscape architects. The Urban Land Institute (ULI) is an association of 40,000 members that represent the entire spectrum of the real estate industry and land-use disciplines. Public Architecture is an allied organization that connects nonprofits with architecture and design professions. And the Landscape Architecture Foundation (LAF) supports the preservation, improvement, and enhancement of the environment.

Three themes have emerged here that relate directly to the capacity organizations: diversity; global perspectives, responsibilities, and exchange; and advocacy and activism.

Nancy Somerville: The landscape architecture profession does not currently look like America and unless we do something to broaden who comes into the profession, we will not be representing the communities that the profession needs to serve in the future. ASLA, working with the President's Council organizations (LAF, ASLA, CLERB, SILA, and LAAB), set the goal that by the year 2025, we reach parity in the profession with where the population was in 2012. That would require us to get to 12 percent African-American, which is about a tenfold increase, and 17 percent Hispanic, which is slightly more than 100-percent increase.

Kathryn Moore: IFLA is 73 nation-states, each one with a delegate on the World Council—a true embodiment of diversity. Within the profession, we may share common values, but must recognize that the practice of landscape architecture is very different in Ethiopia than it is in the United States, in Jordan or Bolivia than it is in New Zealand. We need to support and cultivate a diversity of practice and of educational programs, recognizing that the landscape shapes culture and identity. This ethos underpins the work that many of our delegates are doing to help develop and shape international legislation. We need a far more expressive and interdisciplinary definition of design and landscape and this needs to inform professional and educational documents.

Patrick Phillips: About 85 percent of ULI's 40,000 members are based in the United States. We have had international dimensions to our program dating back to the 1960s, but only recently has it been a matter of following our mission as opposed to just following our members and recognizing that, in particular, the dramatic urbanization that is occurring creates a unique opportunity for ULI to lead. We discovered that we could not rely on the organic growth of a US-based organization with international aspirations; we needed to restructure and redesign the organization. Many of us thought that the greatest challenge toward global growth would be reconciling the cultural distinctions and differences around the world with respect to urban development, real estate practice, planning, design, and so forth. We found that those were pretty manageable and that ULI's approach—multidisciplinary, nonpartisan, nonideological, pragmatic, based on what works—resonated everywhere.

John Peterson: The main program of Public Architecture that is of interest today is "1+," where design professionals give a minimum of 1 percent of their time to pro bono service. We started with architecture, but the vision was always about the designers of the built environment, and landscape architects are a very important part of that group. We are in all 50 states, at over 1,500 firms, doing about $60 million worth of pro bono services annually. We are essentially an online dating site between nonprofit organizations and those that want to give their time to pro bono service. We see the program being used not only to support the social sector, the nonprofit sector, and underserved communities, but also the firms and firm culture, research and development, and different opportunities that become available to firms when you take away the burden of fees.

Stephanie Rolley: Professional societies have long been advocates for policies that further the principles of landscape architecture and advance our profession. Where do you see your organization on the spectrum of advocacy and activism, and do you see a change in your future role as the new generation comes into your organizations?

Nancy Somerville: Advocacy at both the federal and state level is one of ASLA's top priorities. We start with the issues on which the landscape architecture profession speaks with expertise. We bring the members out to show their work and what they can bring to the table for public policy discussions. Issues like active transportation, complete streets, healthy communities, and green infrastructure resonate with public policymakers. Some of our work is pushing the agenda forward, and some is making sure not to lose what we already have.

Barbara Deutsch: One of our key strategies is to do research to provide content that others can use to help make their case. We have to keep going to the mat at the federal level; but the local level is where the mayors are getting together and, for example, establishing their own climate change initiatives. Our role is to help provide the tools to make it easy to call the mayor.

Advocacy is not just lobbying; it is also what I call the activism of the grit—working with the office of planning and getting the right codes in, working with your landscaping committee for your homeowners association—all the day-to-day things that are affecting larger-scale implementation of what we all know and love.

Raquel Peñalosa: A lot of the activism is coming from the younger generation. The connectivity that they have today is making them much more involved and concerned with working with the population, getting involved in putting the word out there, and being extremely creative. We need to create spaces within our organizations for students and emerging professionals and give them a voice and let them inform us.

Barbara Deutsch: We need a new narrative for what landscape architects do, and part of that ties into understanding what is important to people, how they perceive the environment, and what resonates.

Nancy Somerville: We are definitely making progress in public awareness. Our tagline, "landscape architecture: your environment designed," is helpful. But the older you are, the better educated and the wealthier you are, the more likely you are to understand what landscape architecture is. We need to get more involved in career discovery—to reach kids, parents, teachers, and even the ones who are not going to come into the profession. It is a wonderful way to connect with the values that a lot of the young people have that are so attractive.

Advocacy is a long game; it takes focus and persistence and a lot of hands.

Chapter 43

RETROSPECT AND PROSPECT PANEL

PANELISTS

Cornelia Hahn Oberlander, Landscape Architect

Laurie Olin, Partner, OLIN; Practice Professor of Landscape Architecture, University of Pennsylvania School of Design

Peter Walker, Senior Partner, PWP Landscape Architecture

MODERATOR

Frederick "Fritz" Steiner, Henry M. Rockwell Chair in Architecture, Dean, School of Architecture, University of Texas at Austin

The Retrospect and Prospect panel brought together some of the doyens of landscape architecture to talk about where the profession has been and where it is going. The following is a condensed and edited version of the panelists' remarks and is not intended to be a verbatim or complete representation.

Cornelia Oberlander: Landscape architecture is not only a fine art but also a science. Modern design played a large role in the last century. The other very important events in the last century were Rachel Carson's *Silent Spring*, Earth Day, and the Brundtland Report. Published in 1987, the Brundtland Report recognized that environmental problems were global in nature and urged the United Nations General Assembly to establish policies for sustainable urban development. The other milestone is climate change and sustainability. We have to address this in every landscape, every building. The 2015 United Nations Climate Change Conference may have been our last chance for meaningful agreement to shift from fossil fuels to renewable energy before ongoing damage to the world's climate becomes irreversible and devastating. Today more than ever, the design professions can play an important role in ameliorating conflict between the built and natural environment. The scale of environmental challenges demands a new group of multidisciplinary professionals who collaborate to solve our problems.

E. O. Wilson's biophilia hypothesis suggests a biologically based instinctive bond between humans and their environment. In short, longing for nature is built into our genes. He further urges us to keep every scrap of nature in and around our city. Nature holds the key to our aesthetic, intellectual, cognitive, and even spiritual satisfaction.

Peter Walker: We were asked to think a little bit about the key events or people guiding our careers. Stanley White introduced me to the idea that landscape architecture is primarily cultural, that it serves people in that we make things that they need. But it also represents people because the things we make stay and are important to the culture we live in.

Hideo Sasaki introduced us to the idea that the times were changing and that very soon after the war, there was going to be an explosion of schools, land uses, shopping, and transportation. And what he held up as a solution for this was collaboration. At Harvard, almost everybody in those days did collaborative studios. We worked with planners and architects. And we learned how to speak their language and they, to some extent, learned how to speak ours. Sasaki, of course, went on to form two very large collaborative projects. I was part of both of those. And his idea was having collaboration under one roof.

When I was a graduate student, Dan Kiley asked me if I had ever been to Sceaux or the Château de Villandry and told me I had better go. When I went to Europe for the first time and went into those gardens, among others, I realized that Dan was telling me that landscape architecture is an art. And could be a great art. I remember walking Sceaux, thinking this is like Chartres or a cathedral. These are incredible works of art. When you go there, they command your attention in the same way that great art, great architecture, and great music do.

I believe that the artifacts are important. They mark the time. They become the thing you remember about a particular city, about a friend of yours, about something that happened. They are so deep in complex meanings. This is what shaped my thinking over the years.

Laurie Olin: I took longer than usual to invent myself. And for me, each of what might be called milestones occurred over a somewhat extended period of time. When I was eight years old, my family moved to Alaska. It was powerful. Nature was enormously energetic and beautiful. We had fabulous summers and very difficult winters. But there was music, books, art, and stories. And there were also paleontologists, anthropologists, physicists, geologists, engineers, biologists, prospectors, Native Americans, and bush pilots all around me with whom I spent time. And the rivers, the mountains, the volcanoes, earthquakes, the floods. It was great luck to be living there.

I lived, explored, drew, and painted. The takeaway that changed my life—everything comes from nature. We are part of nature. We are in nature all the time, even now right here in this room. All forms come from nature. Whether stochastic or chaotic, entropic, orderly, static, balanced, or dynamic—all that is in nature. And as Buckminster Fuller once said, the opposite of nature is impossible.

The second big watershed moment in my life was when I fell in love with cities. I went to the University of Washington in Seattle, and I had teachers who were profoundly influential—Richard Haag, Vic Steinbrueck, Ted Roethke. Poets, architects, landscape architects. A galaxy of powerful mentors and inspiring artists expanded my horizons. Then I moved to New York, an even bigger city, and it was going 24 hours a day. It was urban life, energy, stimulus. Central Park was full of happenings. There were protest marches. It was a very exciting place.

I realized that cities are landscapes. And that you could add buildings to a landscape, but it did not work the other way around. Landscape was not a sauce you poured over buildings to make them taste good. In a nested order of things, architecture exists within the larger order in the context of landscape the way furniture does in a room.

Then I went to Europe. I lived with painters and sculptors, architects, composers, archeologists, classicists, historians, writers. It was my graduate school. It was the humanities education I felt I needed—a grounding in the history of thought. The knowledge that nothing is really original, that art learns from art, and science learns from science. The purpose of art is simply, I realized, to make old verities and banal things from everyday life—to make them fresh, to find ways of seeing things anew, and get as much pleasure and understanding from them as possible.

Fritz Steiner: What are the prospects for urban-based design moving forward?

Cornelia Oberlander: I think we cannot solve our building and sites unless we collaborate with other professions.

Peter Walker: People are moving back into the cities and there is a market for great things to be done. And that excites me. I have been working all my life to figure out how to be good at (urban) projects, how to make each one really different and memorable. And I was lifted by the possibility that there are still a few more in there of some significance.

Laurie Olin: We may be doing a good job with a few cities in America and Europe, but how about those other millions and billions of people in Africa, Asia, South America, et cetera? We must make living at high density a high-quality experience where people want to raise their kids and it is a great place that is exciting and healthy. We have to make cities great places.

Cornelia Oberlander: We must make them healthy cities.

Laurie Olin: Everybody in the room knows that it is about water and air. And we have to design them so that they have access to natural phenomena, period. We are animals. We need nature; it does not need us.

SPECIAL THANKS

As part of its 50th anniversary activities, the Landscape Architecture Foundation (LAF) launched a capital giving campaign, raising over $4.23 million to expand and enhance its programs. LAF would like to recognize the following donors who contributed to the *LAF: 50 and Forward Campaign* at or above the Advocate level. A full list of donors can be found at www.lafoundation.org.

AECOM

Anonymous—in memory of Debra Mitchell

Anova

Aquatic Design & Engineering

Phillip Arnold

BrightView

Joe Brown and Jacinta McCann

Burton Landscape Architecture Studio

Coldspring

Mark and Doreen Dawson

D'Arcy and Diane Deeks

Design Workshop

Barbara Deutsch

Thomas and Gerry Donnelly

EDSA

Kona Gray

Jennifer Guthrie

Jeanne Lalli

L. M. Scofield Company

Landscape Forms

Landscape Structures

Jim Manskey

Debra Mitchell

Signe Nielsen

OLIN

Ruppert Landscape

Lucinda Reed Sanders

Sasaki

Sally Schauman

SmithgroupJJR

Laura Solano

SWA

TBG Partners

Beth Wehrle

The *New Landscape Declaration* very succinctly lays out the multifaceted social and environmental challenges faced by current, and especially future, generations. Leadership and education will be key in meeting these challenges, and landscape architects are uniquely positioned to lead these efforts. At EDSA, we are proud to support the *New Landscape Declaration* and all that it stands for.

Douglas C. Smith
President, EDSA

As landscape architects, we have direct experience and knowledge to lead and contribute solutions to many pressing environmental and social issues facing our planet. The *New Landscape Declaration* encapsulates the very essence of these times and underscores that we have the opportunity and responsibility to stand up and play a role in making the world a better place for all.

Mark O. Dawson
Managing Principal, Sasaki

Landscape Forms is proud to support the *New Landscape Declaration* and the great work LAF is doing to envision and invest in the transformative power of design. The profession of landscape architecture is dynamic, bold, and creative, and is perfectly poised to be a leader in reshaping our natural and built environment.

Richard Heriford
President, Landscape Forms

We are very proud to support the *New Landscape Declaration*. In countless ways, it mirrors many of the same ideals that have driven and continue to drive Landscape Structures—to create engaging environments where children and their families are encouraged to gather, socialize, play, and exercise, inspiring them to lead meaningful lives with a healthy respect for the world around them, for themselves, and for others.

Steve King
Chairman, Landscape Structures Inc.

My late husband, Joe Lalli, entered the profession at a time when one had to explain that landscape architecture was not just about choosing plants. He would be so proud of the *New Landscape Declaration*, since working across cultures and disciplines was what interested him the most. It was his dedication to the profession and its possibilities that led us to support LAF.

Jeanne Lalli

IMAGE CAPTIONS AND CREDITS

viii *Early Melt on the Greenland Ice Sheet*

Every spring or early summer, the surface of the Greenland Ice Sheet transforms from a vast white landscape of snow and ice to one bejeweled with blue meltwater streams and lakes. In 2016, the transition started early and fast. Surface melt can directly contribute to sea level rise via runoff. It can also force its way through crevasses to the base of a glacier, temporarily speeding up ice flow. Ponding of meltwater can "darken" the ice sheet's surface, which leads to further melting.

NASA Earth Observatory image by Jesse Allen, using Landsat data from the US Geological Survey and EO-1 ALI data provided courtesy of the NASA EO-1 team. Caption adapted from text by Kathryn Hansen. The natural-color image was captured by the Advanced Land Imager (ALI) on NASA's Earth Observing-1 satellite on June 15, 2016.

xii *Teshekpuk Lake, Alaska's North Slope*

Much of Alaska's North Slope is a lacy, lake-dotted expanse of tundra. In January 2006, the US Department of the Interior approved oil and gas drilling on approximately 500,000 acres of land in and around Teshekpuk Lake within the National Petroleum Reserve. Up to 90,000 geese nest in this area, and up to 46,000 caribou use the area for calving and migration.

NASA Earth Observatory image provided by NASA/GSFC/METI/ERSDAC/JAROS and the US/Japan ASTER Science Team. Caption adapted from text by NASA. The picture was captured by the Advanced Spaceborne Thermal Emission and Reflection Radiometer (ASTER) on NASA's Terra satellite on August 15, 2000.

4 *Cultivating Egypt's Desert*

East Oweinat, a land reclamation project in Egypt's hyper-arid Western Desert, creates agricultural fields using irrigation from sprinklers that rotate around a central pivot point. The water comes from the Nubian Sandstone Aquifer, which recharges slowly and is considered a nonrenewable resource.

NASA Earth Observatory image by Joshua Stevens, using Landsat data from the US Geological Survey. Caption adapted from text by Kathryn Hansen. The natural-color image was captured by the Operational Land Imager (OLI) on the Landsat 8 satellite on February 26, 2017.

24 *Meandering in the Amazon*

The Rio Mamoré, flows north (left), receiving a large amount of sediment at the confluence with the Rio Grande (top). The extra sediment enhances the growth of point bars along the inside bends of the riverbank and causes cutoff events, where crescent-shaped oxbow lakes are formed. Sediment and the presence of an erodible river corridor are required to sustain meandering river dynamics.

NASA Earth Observatory image by Jesse Allen, using Landsat data from the US Geological Survey. Caption adapted from text by Kathryn Hansen. The picture was captured by the Thematic Mapper (TM) on the Landsat 5 satellite on June 11, 1985.

40 *Growing Deltas in Atchafalaya Bay*

The delta plain of the Mississippi River has shrunk by nearly 2,000 square miles over the past 80 years. Atchafalaya Bay stands as a notable exception, with new land forming at the mouths of the Wax Lake Outlet and the Atchafalaya River. The key reason is that the Atchafalaya delivers sediment to the coast at a pace that allows it to settle into shallow water and to maintain marshes. In contrast, an extensive series of levees keeps the Lower Mississippi River's water flowing in a narrow channel that whisks water and sediment out to sea.

NASA Earth Observatory image by Joshua Stevens, using Landsat data from the US Geological Survey. Caption adapted from text by Adam Voiland. The false-color image was captured by the Operational Land Imager (OLI) on the Landsat 8 satellite on December 1, 2016.

60 *Checkerboarding in Northern Idaho*

Alongside the Priest River in northern Idaho, the distinctive checkerboard pattern appears to be the result of forest management. White patches reflect areas with younger, smaller trees, while dark green-brown squares are parcels of denser, intact forest. This method maintains some of the forest function while allowing for timber harvesting.

NASA Earth Observatory image provided by the ISS Crew Earth Observations Facility and the Earth Science and Remote Sensing Unit, Johnson Space Center. Caption adapted from text by Andi Hollier, Hx5, and M. Justin Wilkinson, Texas State University, JETS Contract at NASA-JSC. Astronaut photograph ISS050-E-28519 was taken by a member of the Expedition 50 crew on January 4, 2017.

80 *Grand Turk Island*

On the southern end of the Bahamas archipelago, Grand Turk Island is just six miles long, with much of it occupied by Cockburn Town (population about 5,000), the capital of Turks and Caicos. The complex patterns on the east coast (top of the image) are a fraction of a vast reef ecosystem that stretches discontinuously for nearly 600 miles.

NASA Earth Observatory image provided by the ISS Crew Earth Observations Facility and the Earth Science and Remote Sensing Unit, Johnson Space Center. Caption adapted from text by Andi Hollier, Hx5, JETS Contract at NASA-JSC. Astronaut photograph ISS050-E-41317 was taken by a member of the Expedition 50 crew on February 12, 2017.

100 *Drainage Patterns and Wind Farms in Northwest China*

The southwestern end of the Gobi Desert in China's Gansu Province is a desert landscape surrounded by mountains and rolling hills. Snow melt from the higher elevations flows down into streams, forming narrow alluvial fans. The grid pattern superimposed on the basin is part of the Gansu Wind Farm Project where narrow roads mark the paths between dozens of wind turbines.

NASA Earth Observatory image provided by the ISS Crew Earth Observations Facility and the Earth Science and Remote Sensing Unit, Johnson Space Center. Caption adapted from text by Andi Hollier, Hx5, JETS Contract at NASA-JSC. Astronaut photograph ISS050-E-29783 was taken by a member of the Expedition 50 crew on December 29, 2016.

114 *Kangaroo Island Bushfires*

The landscape shows signs of trauma after fires broke out on South Australia's Kangaroo Island in December 2007. In this simulated true-color image, green indicates vegetation, gray-beige indicates bare ground, and charcoal gray indicates the burn scar left behind in Flinders Chase National Park and Ravine des Casoars Wilderness Protection Area.

NASA Earth Observatory image created by Jesse Allen, using data provided by NASA/GSFC/METI/ERSDAC/JAROS and the US/Japan ASTER Science Team. Caption adapted from text by NASA. The picture was captured by the Advanced Spaceborne Thermal Emission and Reflection Radiometer (ASTER) on NASA's Terra satellite on December 20, 2007.

142 *Gillette Coal Pits, Wyoming*

The open pits of several coal mines near the small town of Gillette appear as angular gashes in the snow-covered landscape of northeastern Wyoming. Here, the coal lies at a very shallow depth, making it economical to mine. The steep walls of the overlying rocks cast strong shadows, and wind distributes coal dust so that the pits appear darker than the surrounding landscape.

NASA Earth Observatory image provided by the ISS Crew Earth Observations Facility and the Earth Science and Remote Sensing

Unit, Johnson Space Center. Caption adapted from text by Justin Wilkinson, Texas State University, JETS Contract at NASA-JSC. Astronaut photograph ISS046-E-3395 was taken by a member of the Expedition 46 crew on December 28, 2015

162 *Fish Ponds and Rice Fields, Lower Guadalquivir River*

Fish ponds appear as multicolored geometric shapes on Isla Mayor, an island in the delta of the Guadalquivir River in southwestern Spain. Adjacent to the Doñana National Park marshland reserve, the fish farming practiced here tries to mimic natural conditions. These large ponds are fed with river water, which contains natural food types, without commercial fish feed or antibiotics.

NASA Earth Observatory image provided by the ISS Crew Earth Observations Facility and the Earth Science and Remote Sensing Unit, Johnson Space Center. Caption adapted from text by Justin Wilkinson, Texas State University, JETS Contract at NASA-JSC. Astronaut photograph ISS051-E-12705 was taken by a member of the Expedition 51 crew on April 12, 2017.

170 *Bhutan Himalayas*

The impressive Bhutan Himalayas, one of the highest mountain reliefs on Earth, are permanently capped with snow, which extends down valleys in long glacier tongues. Along with polar ice, mountain glaciers are the features most sensitive to warming temperatures. Because of weather patterns and differences in topography, the glaciers on the southern side of the mountain are stagnating and are more sensitive to climate change than those on the north side.

NASA Earth Observatory image provided by NASA/GSFC/METI/ ERSDAC/JAROS and the US/Japan ASTER Science Team. Caption adapted from text by NASA. The picture was captured by the Advanced Spaceborne Thermal Emission and Reflection Radiometer (ASTER) on NASA's Terra satellite on November 20, 2001.

207 *Paris at Night*

The pattern of Paris' street grid dominates at night. The brightest is the Avenue des Champs-Élysées, the historical axis of the city, which is one of the eleven major boulevards that form a star-like point at the Arc de Triomphe. The thin black line of the winding Seine River is visible with street lights on both banks. The many forested parks of Paris, such as the Bois de Boulogne and Vincennes, stand out as black polygons.

NASA Earth Observatory image provided by the ISS Crew Earth Observations Facility and the Earth Science and Remote Sensing Unit, Johnson Space Center. Caption adapted from text by Justin Wilkinson, Texas State University, JETS Contract at NASA-JSC. Astronaut photograph ISS043-E-93480 was taken by a member of the Expedition 43 crew on April 8, 2015.